RICHARD G. BROWN

transformational geometry

Silver, Burdett & Ginn Inc.

SPECIAL EDITION PUBLISHED BY
DALE SEYMOUR PUBLICATIONS

Printed in the United States of America. Published
simultaneously in Canada.

Order number DS01900
ISBN 0-86651-465-1
(previously published as 0-663-26176-7)

SPECIAL EDITION
PUBLISHED AND DISTRIBUTED BY

DALE
SEYMOUR
PUBLICATIONS
P.O. BOX 10888
PALO ALTO, CA 94303

with the permission of Silver, Burdett & Ginn Inc.   bcdefghi-MA-9043210

Contents

1 The Geometry of Transformations

PAGE

1.1	Maps and Mappings	2
1.2	One-to-One Mappings; Transformations	5
1.3	Mappings in Algebra; Functions	9
1.4	Isometries	12
1.5	Problems Solved by Reflections	17
1.6	Properties of Isometries	22
1.7	Rotations	26
1.8	Translations and Glide Reflections	31
1.9	Symmetry	37
1.10	The Fundamental Theorems of Isometries	41

2 The Algebra of Transformations

2.1	The Composite (Product) of Mappings	46
2.2	The Algebra of Translations	55
2.3	The Algebra of Half-Turns	62
2.4	The Algebra of Rotations	68
2.5	Groups	70
2.6	Transformation Groups	76
2.7	Symmetry Groups	80

Selected Answers	85
Index	91

Preface

This book offers a geometric point of view different from that of the standard course in Euclidean geometry. The study of transformations provides a fresh insight into standard geometric problems and also yields solutions to many other problems too difficult for synthetic or coordinate methods. But, more important, the transformation point of view serves to unify mathematics. The concept of a transformation illuminates the studies of functions, vectors, groups, matrices, complex numbers, and linear algebra. In this book, we shall consider the relationship of transformations to functions, vectors, and groups.

A transformation is a function which students have little trouble understanding, because they actually visualize domains and ranges as geometric sets of points. Also, they can see the tangible effect of a transformation on a geometric figure. Moreover, students find much of the function terminology more meaningful in the geometric context. Words such as mapping, image, fixed point, line symmetry, and point symmetry originated in geometry.

This book may be used over a four- to six-week period with students taking a standard high-school geometry course. It may also be used in a pre-calculus course to offer a broader view of function concepts. This book is also appropriate as a short course for college and junior-college students, and for those preparing to teach mathematics.

I am indebted to The Phillips Exeter Academy for its support of this project and to my colleagues and our students who tested it. I owe a special thanks to Mrs. Irene Little for her secretarial help and to my wife for her unfailing encouragement. To her and our family this book is dedicated.

Richard G. Brown

1

The Geometry of Transformations

1.1 Maps and Mappings

Suppose a space traveler were a very great distance directly above the earth's North Pole. If the skies were cloudless, his view of the earth would look like the map at the top of Figure 1.1. This map was made from a globe of the earth. Every point *P* of the globe's Northern Hemisphere corresponds to exactly one point, *P′*, of the map called the *image* of *P*. This kind of correspondence is often used in making moon maps which show the moon as it is seen from earth. However, this type of map does have a serious disadvantage in that it distorts the shapes of regions near the Equator.

Let us locate on the map the image of the globe's Equator. Figure 1.2 shows that the images of points *P* and *Q* of the globe's Equator are points of the outermost circle of the map. This circle is the image of the globe's Equator. Similarly, the image of the globe's Arctic Circle will be a circle on the map. However, the circular arc from point *Q* to the North Pole, *N*, is mapped to $\overline{Q'N'}$.

Figure 1.1

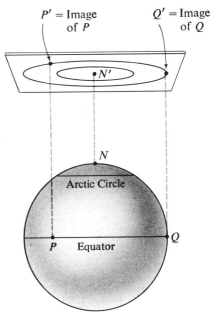

Figure 1.2

Another method for making a map of the Northern Hemisphere is illustrated in Figure 1.3. In this method, the plate of glass has been placed tangent to the globe at its North Pole, and the source of projection is at the globe's South Pole. As in the previous map, the image of the Equator of the globe is the boundary of the map. However, the distortion of regions near the Equator is much less in this map.

Any place on the earth can be located if its latitude and longitude are known. *Circles of latitude,* or *parallels,* are established by passing planes through the earth parallel to the plane of the Equator. The parallels are numbered from the Equator, which has latitude 0, north and south to the poles, which have latitude 90. *Circles of longitude* or *meridians,* are established by planes which pass through the North and South Poles.

There are many methods for making maps, but each method is based on a correspondence between points of the earth's surface and points of a map. The correspondence itself is called a mapping. Each mapping we have discussed has resulted in a map which has had distortions. Later, we shall discuss mappings which produce maps with no distortions. For these mappings, the distance between the images of two points is the same as the distance between the points. Mappings with this property are said to *preserve distance*.

Figure 1.3

EXERCISES

1. On the map in Figure 1.1, locate the images of the circles of latitude and the images of the circles of longitude.

2. In Figure 1.3, what is the ratio of the diameters of the Equator and its image? Are the diameters of the Arctic Circle and its image in the same ratio?

3. Refer to Figure 1.1.

   a. Compare the size of a circle of latitude on the globe with the size of its image.

   b. Compare the distance from the Equator to the North Pole along a meridian of the globe with the corresponding distance on the map.

4. A plane is tangent to a globe at its North Pole, *N*. A point *P* in the Northern Hemisphere is projected on a ray from the globe's center to a point *P'* of the plane.

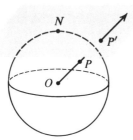

a. What is the image of *N*? (Note. *N* is called a *fixed point*.)

b. What is the image of a circle of latitude?

c. Does the Equator have an image in the plane?

d. Does this mapping seem to preserve distance?

This correspondence is called a Gnomonic mapping. It is used to determine the shortest air or sea route between points of the earth, because every great circle of the globe is mapped onto a straight line.

5. In the figure, a cylinder is tangent to a globe at the Equator. A point *P* of the globe is projected along a ray from the globe's center to its image point *P'* of the cylinder. The map shown illustrates a part of what we would get if the cylinder were opened and laid flat.

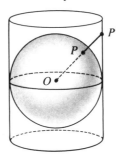

a. Does this mapping preserve distance? Does it preserve distance for points along the Equator?

b. Notice that on this map, the image of Greenland is half the size of the image of South America. On the globe, South America has about 9 times the area of Greenland. For this reason, we say the mapping does not preserve area. Does it appear that the mapping of Exercise 4 preserves area?

6. Each of the six faces of a box is tangent to a globe with center *O*. The top and bottom of the box are tangent to the globe at the North and South Poles. Each point *P* of the globe is mapped to a point *P'* of the box as shown.

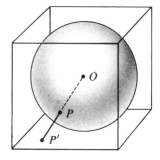

a. Describe the image of the Equator.

b. Describe the image of the Tropic of Cancer.

c. Describe the image of the Arctic Circle.

d. What is the image of the North Pole?

e. Does this mapping preserve distance? area?

f. What regions of the globe will be distorted least by this mapping?

Although this mapping is of no practical importance, it is very much like a mapping of considerable accuracy in which a globe is placed within an icosahedron, a surface with twenty faces, each an equilateral triangle. If the icosahedron is cut and unfolded you get the following map. The given map does not split North and South America. By cutting out and arranging the equilateral triangles in other ways, other areas such as Europe and Asia can be better represented.

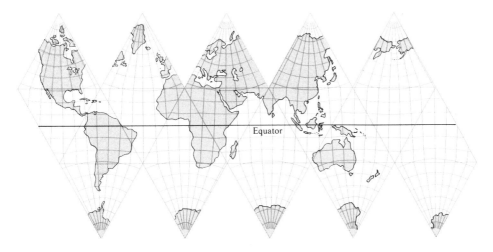

Equator

1.2   One-to-One Mappings; Transformations

In Section 1.1, we discussed mappings from a globe to a plane. For our purposes in plane geometry, we must now turn our attention to mappings between sets of points in a plane.

Definition. If A and B are sets of points in a plane, then a **mapping** M from A to B is a correspondence which associates with each point of A exactly one point of B.

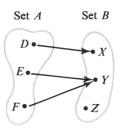

Set A Set B

If M maps point P of set A to point Q of set B, we shall write M:$P \rightarrow Q$. In this case, Q is the image of P, and P is the preimage of Q. Set A is the domain of the mapping. The subset of points of B which are images of points in A is the range of the mapping. In Figure 1.4, M:$D \rightarrow X$, M:$E \rightarrow Y$, and M:$F \rightarrow Y$. X is the image of D, and D is the preimage of X. Z has no preimage in set A; so although Z is in set B, Z is not in the range of M.

Figure 1.4

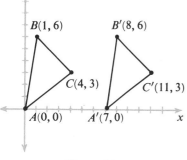

Figure 1.5

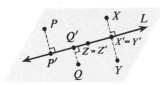

Example 1. Consider point O, segment AB, and line L, shown in Figure 1.5. For each point P of $\overline{AB}$, let there correspond a point P' of L which is the intersection of ray OP and L. This correspondence is a mapping with domain $\overline{AB}$ and range $\overline{A'B'}$.

Example 2. Assume that a given plane has a coordinate system. M is a mapping that corresponds point P with coordinates (x, y) to point P' with coordinates $(x + 7, y)$. We may write this as $\mathsf{M}{:}P(x, y) \rightarrow P'(x + 7, y)$. Thus, M maps every point of the plane seven units to the right. In particular, it maps $\triangle ABC$ to $\triangle A'B'C'$ as shown in Figure 1.6. Since every point of the plane has both an image and a preimage, the entire plane is both the domain and the range of the mapping.

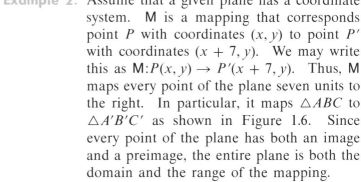

Figure 1.6

Example 3. Consider the mapping that associates with every point P of a plane the foot of the perpendicular from P to a given line, L. Let each point of L be associated with itself. This mapping is called a *projection of the plane* onto L. The domain of the mapping is the plane and the range is the line L.

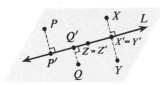

Figure 1.7

Unlike the mapping of Example 3, the mappings of Examples 1 and 2 assign to each point of the range exactly one point of the domain. This distinction leads to the following definition.

Definition. A mapping from a set A to a set B is a **one-to-one mapping** if no two distinct elements of A have the same image in B.

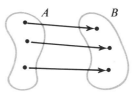 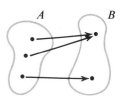

Figure 1.8   A one-to-one mapping   Not a one-to-one mapping

Definition. A **transformation of the plane** is a one-to-one mapping whose domain and range are both the entire plane.

We sometimes refer to a transformation of the plane as simply a transformation. The mapping in Example 2 is a special transformation because it maps every figure to another figure of the same size and shape. The transformation in the next example does not do this.

Example 4. Consider the mapping defined as follows: $M:P(x, y) \rightarrow P'(x + 7, 2y)$. Recall that this means that M maps point P with coordinates (x, y) to the point P' with coordinates $(x + 7, 2y)$. To find the image of $\triangle ABC$ we plot the images of several points of $\triangle ABC$ as shown in Figure 1.9. Later we shall be able to prove that all the points of $\triangle ABC$ get mapped to $\triangle A'B'C'$. Because M is one-to-one and has the entire plane as its domain and range, it is a transformation.

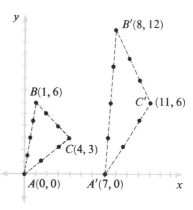

Figure 1.9

EXERCISES

1. O is a point midway between the parallel lines L_1 and L_2. A mapping M associates with each point P of L_1 the point P' which is the intersection of ray PO and L_2.

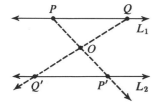

a. What are the domain and range of M?

b. Is M a one-to-one mapping?

c. Does M preserve distance?

2. O is a point in a plane. With every point P of the plane, the mapping M associates a point P', so that P is the midpoint of $\overline{OP'}$. The image of O is itself.

a. Draw a diagram which shows O, P, and P'. Select a point Q, and locate Q', its image by M.

b. Can you draw any conclusion about the distance from P to Q and the distance from P' to Q'?

c. What are the range and domain of M?

d. Is M a one-to-one mapping?

e. Is M a transformation?

3. S_1 and S_2 are squares. The mapping M associates with each point P of S_1 the point P' which is the intersection of ray OP and S_2.

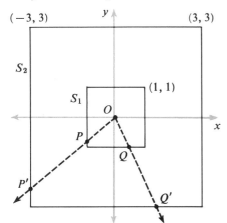

a. Find the images of $(-1, 0)$ and $\left(\frac{1}{2}, 1\right)$.

b. Find the preimages of $(2, 3)$ and (a, b).

c. Suppose P and Q are any two points of the domain with images P' and Q'. Compare the distance from P to Q with that from P' to Q'.

4. a. Let $ABCD$ be a trapezoid with bases $\overline{BC}$ and $\overline{AD}$. Describe a one-to-one mapping between $\overline{AB}$ and $\overline{CD}$.

b. Describe a one-to-one mapping between two parallel lines.

5. Consider the mapping M such that $\mathsf{M}{:}P(x, y) \to P'(y, x)$. Points $A(0, 0)$, $B(5, 1)$, and $C(6, 3)$ are the vertices of a triangle. Draw $\triangle ABC$ and find its image by plotting the images of several points of the triangle. Does M preserve distance?

6. Consider the mapping M such that $\mathsf{M}{:}P(x,y) \to P'(x, 2y)$. Points $A(-1, 3)$, $B(4, 1)$, and $C(2, 0)$ are the vertices of a triangle. Draw $\triangle ABC$ and find its image by plotting the images of several points of the triangle. Does M preserve distance? Does M preserve the shape of $\triangle ABC$?

7. Repeat Exercise 6 for the mapping $\mathsf{M}{:}P(x, y) \to P'(2x, 2y)$.

8. Repeat Exercise 6 for the mapping $\mathsf{M}{:}P(x, y) \to P'(-x, -y)$.

9. Consider a projection of the plane onto line L.

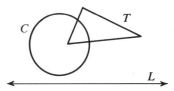

a. Describe the image of the circle, C, and the image of the triangle, T, shown in the given figure.

b. Describe the image of the union of C and T. Is it the same as the union of the images of C and T?

c. Describe the image of the intersection of C and T. Is it the same as the intersection of the images of C and T?

d. Make a sketch which shows two nonintersecting sets whose images by this mapping do intersect.

(Note. Parts c and d illustrate the fact that this mapping does not preserve intersections. Only one-to-one mappings preserve intersections. All mappings preserve unions.)

1.3   Mappings in Algebra; Functions

The idea of a mapping between sets seems very natural in geometry. Yet, this idea is widely used throughout mathematics and no doubt you have already encountered it in algebra. Perhaps you used the word *function* instead of the word "mapping." The only difference is that mappings in algebra are usually between sets of numbers while the mappings we have discussed in geometry are between sets of points.

Example 1. The correspondence $F:x \rightarrow 2x$, which associates every integer x with the integer $2x$, is a mapping or function from I to I, where $I = \{\ldots, -3, -2, -1, 0, 1, 2, 3, \ldots\}$. The domain of this mapping (function) is I, and the range is the set of even integers. The function is one-to-one.

In an algebra course we might describe the function in Example 1 by writing $F(x) = 2x$. We read this as "F at x equals $2x$" or "F of x equals $2x$." This is a brief way to indicate that the value of the function F, at x, is twice x. Similarly, F(5) is read "F at 5" or "F of 5." In this case, the value of F at 5 is 10 and we write $F(5) = 10$.

Example 2. The function $F:x \rightarrow 3x^2$ can also be written as $F(x) = 3x^2$, and we say that the value of F at x is $3x^2$. In this case, $F(1) = 3$ and $F(2) = 12$. If the domain of this function is the set of real numbers, the range will consist of the set of nonnegative numbers, since x^2 is never a negative number. Morever, this function is not one-to-one since two numbers of the domain, such as 1 and -1, can have the same image, 3.

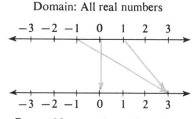

Domain: All real numbers

Range: Nonnegative real numbers

Figure 1.10

In all our examples of mappings thus far, the domain and range have both been either sets of numbers or sets of points. Actually, the word "mapping" is used more generally in mathematics for any correspondence between two sets, provided each element of the domain corresponds with exactly one element of the range. The correspondence that associates each person with his birthday is a mapping whose domain is the set of all persons and whose range is the days of the year. This mapping is not one-to-one, since two people may have the same birthday. If we let **B** denote this mapping, we may write **B**:$x \rightarrow$ birthday of x. This is analogous to writing **F**:$x \rightarrow 3x^2$, in Example 2. Similarly, **B** (Abraham Lincoln) = February 12 is analogous to **F**(1) = 3.

Example 3. The correspondence **M**, which assigns to each angle its degree measure, is a mapping whose domain is the set of all angles and whose range is the set of real numbers between 0 and 180. This mapping is not one-to-one. Can you explain why?

One of the most important mappings in mathematics is the correspondence which associates the points of a plane with their coordinates. The domain of this mapping is the set of all points of the plane. The range is the set of all ordered pairs of real numbers. Let us call this mapping **C**. Of course, there is also a mapping which associates with every ordered pair of real numbers a point of the plane. This mapping is called the *inverse* of **C** and is denoted **C**⁻¹.

$$\textbf{C}:\text{Points of the plane} \rightarrow \text{Ordered pairs of real numbers}$$
$$\textbf{C}^{-1}:\text{Ordered pairs of real numbers} \rightarrow \text{Points of the plane}$$

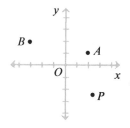

Figure 1.11

$$\textbf{C}:A \rightarrow (2, 1) \qquad \textbf{C}:B \rightarrow (-3, 2) \qquad \textbf{C}:P \rightarrow (x, y)$$
$$\textbf{C}^{-1}:(2, 1) \rightarrow A \qquad \textbf{C}^{-1}:(-3, 2) \rightarrow B \qquad \textbf{C}^{-1}:(x, y) \rightarrow P$$

The mappings **C** and **C**⁻¹ are so familiar to us that we often make no distinction between a point and its coordinates. For example, in Figure 1.11 we may call (2, 1) the point A. Also, rather than refer to point B, we may refer to the point $(-3, 2)$.

Example 4. Consider the mappings **M** and **D** defined as follows.

$$\textbf{M}:(x, y) \rightarrow (-x, y) \qquad \textbf{D}:(x, y) \rightarrow \sqrt{x^2 + y^2}$$

For $P(5, 12)$, $\textbf{M}(P) = (-5, 12)$ and $\textbf{D}(P) = \sqrt{5^2 + 12^2} = 13$. The domain of both mappings is the coordinate plane. However, their ranges are very different. **M** maps every point to another point, and its range is the entire plane. On the other hand, **D** maps every point to a number which equals the distance of the point from the origin. Hence, the range of **D** is the set of nonnegative real numbers.

EXERCISES

1. I is the set of all integers, and F is a function or mapping, with domain I.
   a. $F(x) = 2x + 1$. Evaluate $F(-1)$, $F(0)$, $F(1)$, and $F(2)$. What is the range of F? Is F one to-one?
   b. $F(x) = 5x$. Evaluate $F(-1)$, $F(0)$, $F(1)$, and $F(2)$. What is the range of F? Is F one-to-one?
   c. $F(x) = x + 7$. What is the range of F? Is F one-to-one?

2. F is a function or mapping, whose domain is the set of all real numbers, and $F(x) = 3x^2 + 5$.
   a. Evaluate $F(0)$, $F(1)$, $F(-1)$, and $F(2)$.
   b. What is the range of F?
   c. Is F one-to-one?
   d. Find the preimages of 305 and 26.

3. Give an example of a mapping between each pair of sets.
   a. A set of people to a set of people
   b. A set of people to a set of numbers
   c. Is either mapping one-to-one?

4. M is a mapping which associates every nonvertical line with its slope.
   a. Why is the word "nonvertical" necessary?
   b. What is the range of this mapping?
   c. Is M one-to-one?

5. The correspondence $F: x \to \frac{1}{2}x$ is not a function from I to I, but is a function from R to R. Explain why. (I denotes the set of integers, and R denotes the set of real numbers.)

6. D associates each point of the plane with its distance from the origin.
   a. P is the point $(3, 4)$. Evaluate $D(P)$.
   b. Find another point, Q, for which $D(Q) = D(P)$.

7. The mapping $M:(x, y) \to (2x, y)$ is defined for all points of the plane.
   a. What are the images of $(5, -3)$, and $(0, 2)$?
   b. What are the preimages of $(7, 1)$, $(0, -5)$, and (a, b)?
   c. Find the range and domain of M.
   d. Is M one-to-one? Is it a transformation?
   e. Give an example of two points, P and Q, and their images, P' and Q', such that the distance from P to Q is equal to the distance from P' to Q'. Does your example prove that this mapping preserves distance?

8. The mapping M is defined for all points (x, y) of the plane as follows.
   If $x \geq 0$, then $M:(x, y) \to (x + 1, y)$.
   If $x < 0$, then $M:(x, y) \to (x - 1, y)$.
   a. Find the images of $(3, -8)$, $(-2, 4)$ and $(0, 0)$.
   b. Find the range and domain of M.
   c. Is M one-to-one? Is it a transformation?
   d. Give an example of two points, P and Q, and their images, P' and Q', such that the distance from P to Q is equal to the distance from P' to Q'. Does your example prove that this mapping preserves distance?

1.4   Isometries

In many of the previous examples and exercises, we have been concerned with geometric mappings that preserve distance. Such mappings have the property that the distance between the images of *any* two points is the same as the distance between the points. Transformations that preserve distance are given a special name.

> **Definition.** An **isometry** is a transformation that preserves distance.

We shall use the symbol PQ to represent the distance from P to Q, or the length of $\overline{PQ}$. Thus, if an isometry maps points P and Q to points P' and Q', then $PQ = P'Q'$ for all P and Q of the plane. We shall now consider two very important types of isometries.

> **Definition.** Let L be a line in a plane. A **reflection in line L** is a transformation that maps every point P of the plane to an image point P' in the following way.
>
> 1. If P is in L, then $P = P'$.
> 2. If P is not in L, then L is the perpendicular bisector of $\overline{PP'}$.

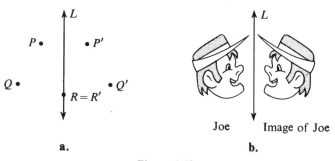

a.   b.

Joe   Image of Joe

Figure 1.12

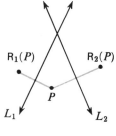

Figure 1.13

Figure 1.12b suggests that line L acts like a mirror. This is why we use the words "reflection" and "mirror image." Line L is called the line of reflection or mirror line. We shall use the symbol R_L to denote a reflection in line L. To distinguish between the reflections in line L_1 and line L_2, we shall use the symbols R_1 and R_2, respectively. To denote the image of point P by a reflection in L_1 we shall use the symbol $\mathsf{R}_1(P)$. Likewise, $\mathsf{R}_2(P)$ shall denote the image of P by a reflection in L_2. Finally, we should note from the definition of a line reflection that if a line reflection maps point P to an image point P', then it also maps P' to P.

The previous definition of line reflection is very similar to the following definition. Carefully compare these two definitions.

Definition. Let O be a point in a plane. A **reflection in point O** is a transformation that maps every point P of the plane to an image point P' in the following way.

    1. If P is point O, then $P = P'$.

    2. If P is not point O, then O is the midpoint of $\overline{PP'}$. (See Figure 1.14.)

       a.   **b.**

Figure 1.14

To find the image of a figure by a point reflection, we could trace the figure on a piece of paper that has been pinned to the plane at point O. If the paper is then turned 180° about point O, the figure on the paper will be the image of the original figure. For this reason, a point reflection is often called a half-turn. We shall denote a half-turn about point O by the symbol H_O. The image of point P by a half-turn about point O is denoted by $\mathsf{H}_O(P)$. (See Figure 1.15.)

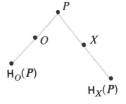

Figure 1.15

Example. Figure 1.16 shows the image of point $P\,(2, 3)$ when reflected in the x-axis, the y-axis, the line L with equation $y = x$, and the origin.

$$\mathsf{R}_x(P) = (2, -3)$$
$$\mathsf{R}_y(P) = (-2, 3)$$
$$\mathsf{R}_L(P) = (3, 2)$$
$$\mathsf{H}_O(P) = (-2, -3)$$

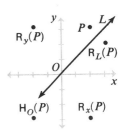

Figure 1.16

It should be clear from the definitions of line reflection and point reflection that the domain and range of each of these mappings is the entire plane, and that each mapping is one-to-one. Thus, each mapping is a transformation. To show each is an isometry we must show $PQ = P'Q'$, where P and Q are any two points of the plane with images P' and Q'.

To show that a line reflection preserves distance, each of the following cases must be considered. In Case a, the most general case, congruent triangles are used to prove $PQ = P'Q'$. The remaining cases are easy. You will be asked to prove them in the exercises.

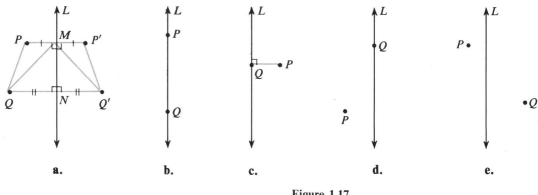

Figure 1.17

To show that a point reflection preserves distance, the following three cases must be considered. In each case, it is easy to show that $PQ = P'Q'$. You will be asked to do this in the exercises. Just remember that O is the midpoint of $\overline{PP'}$ and $\overline{QQ'}$.

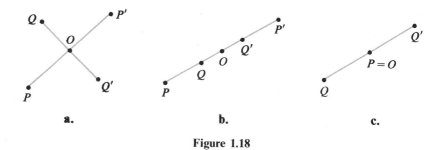

Figure 1.18

We may now state the following theorem.

THEOREM 1. Line reflections and point reflections (or half-turns) are isometries.

EXERCISES

1. Copy the following figures, and find the image of each by a reflection in line *L*.

a.

b.

HELP!
I AM
TRAPPED
BEHIND
THE
LINE!

2. Point *A'* is the reflection image of *A*. Copy the figure, draw the mirror line, *L*, and locate $R_L(B)$ and $R_L(C)$.

• *A'*

• *A*

• *B*

C •

3. Copy the following figures and find their images by a half-turn about point *O*.

a.

b. SWIMS

O

O

4. a. Palindromes are words that spell the same backward and forward, such as RADAR and MOM. When reflected in a vertical line, RADAR becomes ЯAƆAЯ but MOM remains MOM! Can you think of any other words that reflect like MOM?

b. When reflected in a horizontal line, MOM becomes WOW, but BOB remains BOB. Are there any other words that reflect in a horizontal line, like BOB?

c. Spell these sentences backwards.

ABLE WAS I ERE I SAW ELBA
WAS IT A RAT I SAW
RISE TO VOTE SIR

5. People in distress sometimes write SOS in the snow or sand. Why is this signal particularly effective for searching aircraft? Does SOS remain unchanged when reflected in a line? What happens when it is reflected in a point?

6. The given figure shows ink drops on a piece of paper. An ink blot will form if the paper is folded along the dashed line. Sketch this ink blot.

7. a. Show that the line $y = x$ is the perpendicular bisector of the segments joining each of the following pairs of points.

(3, 0) and (0, 3); (3, 4) and (4, 3); (*a*, *b*) and (*b*, *a*)

b. Explain why a reflection in the line $y = x$ is defined by the mapping $M:(a, b) \rightarrow (b, a)$.

c. What mapping defines a reflection in the line $y = -x$?

8. Let L be the line $y = x$. Find the images of each of the following points when reflected in the x-axis, the y-axis, the line L, and the origin.

    a. $P(5, 2)$ **b.** $Q(-2, 0)$ **c.** $R(1, \sqrt{6})$

9. a. O is the origin and P is the point (a, b). What is $H_O(P)$?

    b. Show that you get the same image point by reflecting P in the x-axis, and then reflecting the image in the y-axis.

10. The vertical line L_1 and the horizontal line L_2 intersect at $C(2, 3)$. Give the coordinates of each of the following.

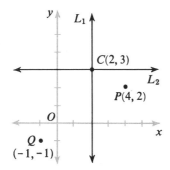

    a. $R_1(P)$ **c.** $R_1(Q)$ **e.** $H_C(P)$
    b. $R_2(P)$ **d.** $R_2(Q)$ **f.** $H_C(Q)$

11. For each of the following pairs of points, determine the equation of L, the line of reflection, so that $P' = R_L(P)$.

    a. $P(1, 1)$, $P'(-1, -1)$
    b. $P(2, 6)$, $P'(4, 8)$

12. For each of the following pairs of points, determine the coordinates of point O such that $H_O(P) = P'$.

    a. $P(0, 0)$, $P'(4, 6)$
    b. $P(-1, 2)$, $P'(1, 4)$
    c. $P(x_1, y_1)$, $P'(x_2, y_2)$

13. Line L has equation $x + y = 4$, and point O is the origin.

    a. What are the coordinates of $R_L(O)$?

    b. What is the equation of $H_O(L)$?

14. Prove that $PQ = P'Q'$ for each case shown in Figure 1.17.

15. Prove that $PQ = P'Q'$ for each case shown in Figure 1.18.

16. a. M is the following transformation: $M:(x, y) \rightarrow (x + 3, y)$. Prove that M is an isometry by letting $P(x_1, y_1)$ and $Q(x_2, y_2)$ be arbitrary points and then using the distance formula to show $PQ = P'Q'$.

    b. If $M:(x, y) \rightarrow (x + 3, y - 2)$, is M an isometry?

17. Given intersecting lines L_1 and L_2, a transformation M maps each point of L_1 to itself. Any other point, P, is mapped to a point, P', so that $\overline{PP'}$ is parallel to L_2 and bisected by L_1.

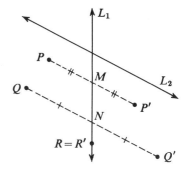

    a. Is M an isometry?

    b. Prove that the area of $\triangle PQR$ equals the area of $\triangle P'Q'R'$.

    c. Is the midpoint of $\overline{PQ}$ mapped to the midpoint of $\overline{P'Q'}$? Try to prove your answer.

1.5   Problems Solved by Reflections

When a ray of light is reflected by a mirror, the measure of the *angle of incidence* equals the measure of the *angle of reflection*. In Figure 1.19, therefore, $m\angle 1 = m\angle 2$. The same is true of balls bouncing off a wall. However, if the ball has a lot of spin, the angles of incidence and reflection may differ slightly.

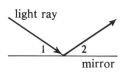

light ray

mirror

Figure 1.19

Suppose you want to roll a ball from A to B by bouncing it off the wall, represented by line L, shown in Figure 1.20. For what point on the wall should you aim? Let $B' = \mathsf{R}_L(B)$ and aim for the point at which L intersects $\overline{AB'}$. The reason this works is because $m\angle 3 = m\angle 2$ by the reflection, and since $m\angle 3 = m\angle 1$, then $m\angle 2 = m\angle 1$. Thus, the angle of incidence is congruent to the angle of reflection. Incidentally, this path is the shortest path from A to B via L because $AP + PB = AP + PB' = AB'$ and the shortest distance from A to B' is along the line containing A and B'.

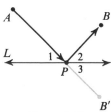

Figure 1.20

Example 1. What is the shortest distance from $A(0, 5)$ to $B(8, 1)$ via the x-axis?

Solution. The reflection image of B in the x-axis is $B'(8, -1)$. (See Figure 1.21) Now, $AP + PB = AB' = \sqrt{(8 - 0)^2 + (-1 - 5)^2} = 10$.

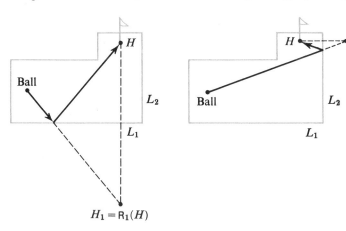

Figure 1.22 shows a par 2 hole on a miniature golf course. To get a hole-in-one, aim for either $H_1 = \mathsf{R}_1(H)$ or $H_2 = \mathsf{R}_2(H)$.

Figure 1.21

Figure 1.22

Of course, if you want to be really tricky, you can bounce the ball off two or three walls, as shown in Figure 1.23.

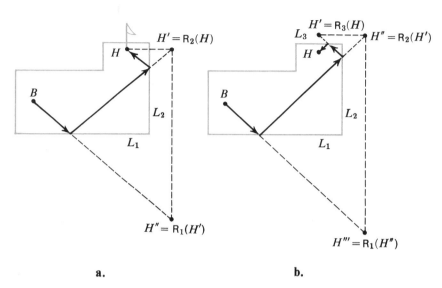

a.　　　　　　　　　　　　b.

Figure 1.23

Of all the paths from B to L_1 to L_2 to H, the shortest path is shown in Figure 1.23a. The total distance traveled by the ball is equal to BH''. Similarly, Figure 1.23b shows the shortest path from B to H via L_1, L_2 and L_3, in that order. Here, the total distance traveled by the ball is equal to BH'''.

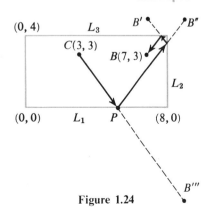

Figure 1.24

Example 2. A 4-foot by 8-foot billiard table is shown in Figure 1.24, with a cue ball at the point, $C(3, 3)$. For what point should you aim in order to hit three sides of the table and then the ball at $B(7, 3)$? How far does the cue ball travel before hitting the other ball?

Solution. If you are to bounce the cue ball off sides L_1, L_2, and L_3, in that order, consider the successive reflection images of B in L_3, L_2, and L_1.

$$B' = \mathsf{R}_3(B) = (7, 5)$$
$$B'' = \mathsf{R}_2(B') = (9, 5)$$
$$B''' = \mathsf{R}_1(B'') = (9, -5)$$

Then, aim the cue ball for B'''. In order to find the coordinates of P, the point of intersection of $\overline{CB'''}$ and L_1, we must first find the equation of line CB'''. Since $\overleftrightarrow{CB'''}$ contains $(3, 3)$ and $(9, -5)$, its slope is $\dfrac{-5 - 3}{9 - 3}$, or $-\dfrac{4}{3}$. The line which has slope $-\dfrac{4}{3}$ and contains the point $(3, 3)$ has the following equation.

$$y = -\frac{4}{3}x + 7$$

When $y = 0$, the value of x is $\dfrac{21}{4}$. Therefore, P has coordinate $\left(\dfrac{21}{4}, 0\right)$. The total distance traveled by the cue ball equals the distance from C to B'''.

$$CB''' = \sqrt{(3 - 9)^2 + (3 + 5)^2} = 10$$

Our final examples are of problems that may be solved by using a half-turn or a line reflection.

Example 3. Given point P, line L, and circle C, as shown in Figure 1.25, find a segment AB, with A in L, B in C, and with P as its midpoint.

Solution. We can find the image of L by a half-turn about P. The image of L will intersect C in two points. Call either of these points B. Let point A be the preimage of B. The midpoint of $\overline{AB}$ is P, so all the given conditions have been fulfilled.

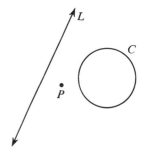

Figure 1.25

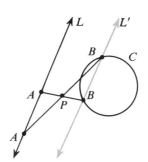

Figure 1.26

Example 4. Given the circles C_1 and C_2 and line L, find points A and B, one a point of each circle, such that L is the perpendicular bisector of $\overline{AB}$.

Solution. We can find the image of C_1 by a reflection in line L. The image of C_1 will intersect C_2 in two points. Call either of these points, B. Point A is the preimage of B. $\overline{AB}$ fills all the given conditions.

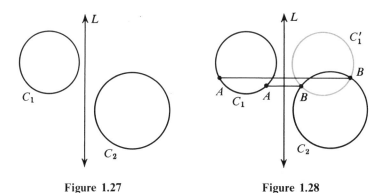

Figure 1.27   Figure 1.28

EXERCISES

1. Show how to score a hole-in-one on this fifth hole of a miniature golf course in each of the following cases. In each case there is more than one possible way.

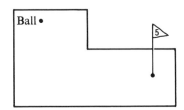

  a. Bouncing the ball off one wall

  b. Bouncing the ball off two walls

  c. Bouncing the ball off three walls

2. Two cities, located at A and B, need to pipe water from the river, represented by line L. They decide to build just one pumping station. Where along the river should they locate it to use the minimum amount of pipe?

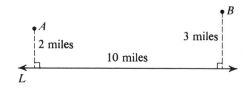

3. Prove that the total length of the golf-ball paths in Figures 1.23a and b are BH'' and BH''', respectively.

4. A ray of light is reflected by two perpendicular mirrors. Prove that the emerging ray is parallel to the initial incoming ray.

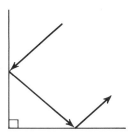

5. A ball rolls at a 45-degree angle away from one side of a 4-foot by 8-foot pool table. If the ball starts at the point $P(0, 3)$, as shown, then it will eventually return to its starting point. Does this happen for any other starting points $P(0, y)$ on $\overline{OD}$?

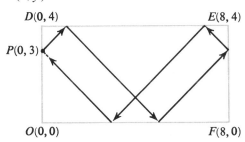

6. Given a triangle, a line, and the point P, use a half-turn to find all line segments, $\overline{AB}$, having P as a midpoint with A in the triangle and B in the line.

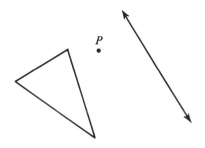

7. Find a line through P which intersects the two circles, C_1 and C_2, in chords of the same length.

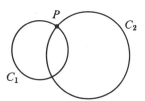

8. A pool shark wants to shoot the cue ball, C, located at the point $(2, 3)$, so that it hits the eight ball $B(6, 1)$. He can do this by bouncing his cue ball off one, two, or three sides, as shown. Find the coordinates of the points X, Y and Z, at which he must aim to make the three shots.

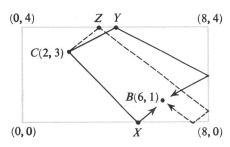

9. Lines L_1 and L_2 are parallel as shown in the figure below.

a. Find the shortest path from A to B which first touches L_2, and then L_1. Show that this path is 20 units long.

b. Show that the shortest path from A to B which touches L_1 first and then L_2, is $8\sqrt{5}$ units long.

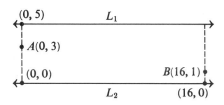

10. Given two circles, C_1 and C_2, and line L, use one of the transformations studied in this section to find a square, $PQRS$, with P and R in L, Q in C_1, and S in C_2. How many such squares are there?

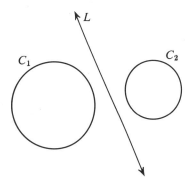

11. Point P in line L has the property that tangents to circles C_1 and C_2 from P form equal angles with L. Use a reflection in line L to find three other points of L which have this property.

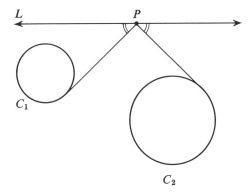

1.6 Properties of Isometries

The sketch in Figure 1.29 illustrates a person and his reflection in a fun house mirror. The image is badly distorted, which, of course, is what makes the mirror so much fun. Long slender legs look stubby and bent; heads and stomachs bulge.

A reflection in an ordinary mirror has none of these properties. We have already stated that a reflection is an isometry, and therefore it preserves distance. Our experience tells us that the reflection images of our straight legs would also be straight, the angle at our elbow would be the same as the angle at our image's elbow, and the parallel stripes of our sweater would look parallel on our image's sweater. In mathematical language, we could say that a reflection preserves straight lines, angle measure and parallelism. In fact, these are properties not only of a reflection, but of every isometry.

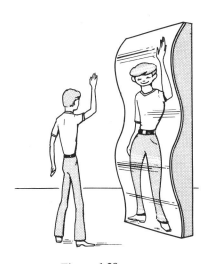

Figure 1.29

THEOREM 2. An isometry preserves collinearity.

> *Proof.* Given three collinear points, A, B, and C, we must show that their images, A', B', and C' are collinear. Recall that point B is between points A and C if and only if $AB + BC = AC$.
>
> If A, B, and C are collinear, one of the points is between the other two. Suppose this point is B. Then, $AB + BC = AC$ by the definition of betweenness. Since each of the distances, AB, BC, and AC is preserved by the isometry, $A'B' + B'C' = A'C'$, and so B' is between A' and C'. Thus, A', B', and C' are collinear, and the theorem is proved.

COROLLARY to THEOREM 2. An isometry maps lines to lines, segments to segments, rays to rays, and angles to angles.

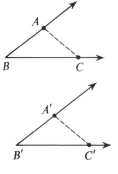

The next theorem states that an isometry maps an angle to an angle which is the same size as the original angle.

THEOREM 3. An isometry preserves angle measure.

> *Proof.* Consider $\angle ABC$ and its image, $\angle A'B'C'$. Since an isometry preserves distance, $AB = A'B'$, $BC = B'C'$, and $AC = A'C'$. Therefore, $\triangle ABC \cong \triangle A'B'C'$, and $m\angle ABC = m\angle A'B'C'$. Thus, the theorem is proved.

Figure 1.30

THEOREM 4. An isometry preserves parallelism.

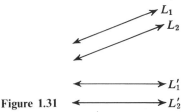

Figure 1.31

> *Proof.* Given parallel lines L_1 and L_2, we must show that their images, L_1' and L_2' are also parallel. We shall use the fact that parallel lines L_1 and L_2 are everywhere the same distance apart. Since the isometry preserves this distance, L_1' and L_2' must also be everywhere the same distance apart. Thus, L_1' and L_2' are parallel.

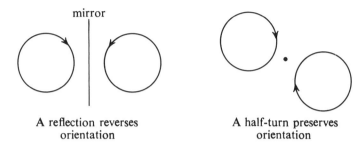

A reflection reverses
orientation

A half-turn preserves
orientation

Figure 1.32

Isometries do not preserve all properties of geometric figures. When you look into a mirror, your right hand is your image's left hand. If you rotate your arm clockwise, your image's arm rotates counter-clockwise. Because of this, we say that a reflection does not preserve orientation. However, a half-turn does preserve orientation.

EXERCISES

1. A' and B' are the images of A and B by an isometry. Explain why the midpoint of $\overline{AB}$ is mapped to the midpoint of $\overline{A'B'}$.

2. Explain why an isometry preserves perpendicularity.

3. Consider the mapping M such that M:$(x, y) \rightarrow (2x, 2y)$.
   a. Does M preserve collinearity?
   b. Does M preserve angle measure?
   c. Does M preserve parallelism? (Note. Parallelism is preserved if the images of any two parallel lines are also parallel.)
   d. Compare the distance between any two points with the distance between their images.

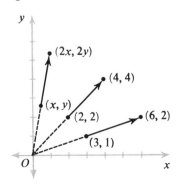

4. Consider the following transformation: $M:(x, y) \rightarrow (x, 2y)$.

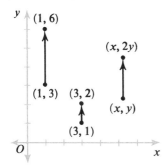

a. Is M an isometry?

b. Does M preserve collinearity? angle measure? parallelism? orientation?

c. Prove that the midpoint of any segment is mapped to the midpoint of the image segment.

5. a. A figure, F, is reflected in a line L_1 and its image F' is reflected in another line, L_2, giving a final image F''. Do F and F'' have the same or reverse orientation?

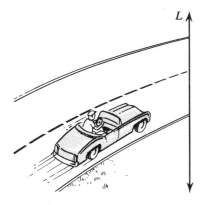

b. The sketch shows an American car on a U. S. highway. Explain why its reflection image in line L looks like an English car on a British road.

6. An isometry preserves the intersection of sets. Using this fact, write another proof of Theorem 4.

7. An isometry maps points A, B, and C to points A', B', and C'. Copy the figure, and using only a compass, locate point C' in each of these cases.

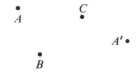

a. The isometry preserves orientation.

b. The isometry reverses orientation.

8. The given figure shows the edge of a fun house mirror. To determine the image of P, find the perpendicular to the mirror from P. The point at which this perpendicular intersects the mirror will be the midpoint of $\overline{PP'}$. Note that a perpendicular to a curved mirror is perpendicular to a tangent line of the mirror. By plotting other image points, show that the image of $\overline{PQ}$ is not $\overline{P'Q'}$.

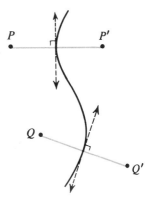

Edge of a fun house mirror

1.7 Rotations

Record players, clocks, and Ferris wheels intuitively suggest the next transformation we shall consider. If a record player were turned on for a second or two, each point of the turntable would rotate to a new position. Mathematically speaking, the position of a point when the record player is turned off is the image of the initial position of the point. We call this type of transformation a rotation. A rotation may be in the clockwise or counterclockwise direction. It is customary to consider a counterclockwise rotation as being in the positive direction and a clockwise rotation as being in the negative direction.

In your geometry course, an angle was probably considered to be a set of points, and its measure in degrees was a number between 0 and 180. We shall use the word "angle" in a more general sense. In Figure 1.33e, point P has been rotated through 380 degrees. Since the magnitude of the rotation is 380°, we say "the angle of rotation is 380°." We shall use the symbol $\measuredangle POP'$ to denote the angle of the rotation about point O which maps P to P'. Therefore, for Figure 1.33e we shall write $\measuredangle POP' = 380$, even though $\angle POP'$ has measure 20. In Figure 1.33a, $\measuredangle POP' = 90$, and $m\angle POP' = 90$, but in Figure 1.33c, $\measuredangle POP' = 270$ and $m\angle POP' = 90$.

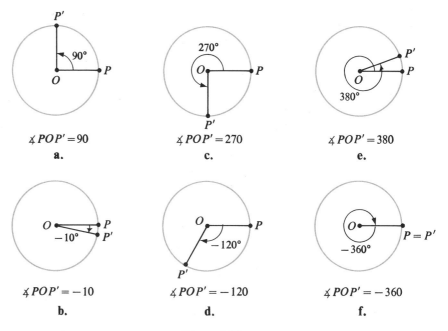

$$\measuredangle POP' = 90$$
a.

$$\measuredangle POP' = 270$$
c.

$$\measuredangle POP' = 380$$
e.

$$\measuredangle POP' = -10$$
b.

$$\measuredangle POP' = -120$$
d.

$$\measuredangle POP' = -360$$
f.

Figure 1.33

Definition. Consider a point P and a real number x. The **rotation of $x°$ with center C**, denoted C_x, is the transformation obtained as follows.

1. The image of C is C.
2. The image of any other point, P, is the point, P', of the circle with center C and radius CP such that $\angle PCP' = x$.

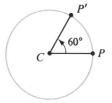
The rotation C_{60}

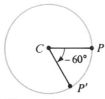
The rotation C_{-60}

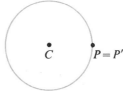
The zero rotation C_0

Figure 1.34

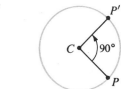

If $x < 0$, the rotation is in the clockwise direction.

If $x > 0$, the rotation is in the counterclockwise direction.

If $x = 0$, every point is its own image, and this zero rotation is called the *identity*.

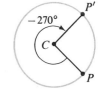

Different symbols may represent the same rotation. For example, C_{90}, C_{-270}, and C_{450} all describe the same correspondence, $P \to P'$. (See Figure 1.35.) What really matters is the final correspondence, not how the rotation is made. For this reason, we write $C_{90} = C_{-270} = C_{450}$, because all three rotations describe the same correspondence between points.

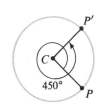

Figure 1.35

THEOREM 5. A rotation is an isometry.

Proof. Let P' and Q' be the rotation images of any two points, P and Q. By the definition of rotation, $\angle PCP' = \angle QCQ'$. Therefore, $m\angle PCP' = m\angle QCQ'$ and since $m\angle PCP' + m\angle PCQ' = m\angle QCQ' + m\angle PCQ'$, then $m\angle P'CQ' = m\angle PCQ$. Also, $CP = CP'$ and $CQ = CQ'$, so $\triangle PCQ \cong \triangle P'CQ'$. Therefore, $PQ = P'Q'$. Since the rotation preserves distance, it is an isometry.

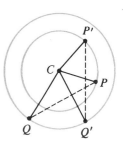

Figure 1.36

Suppose you wanted to trace $\triangle PCQ$ shown in Figure 1.36. If you traced from P to C to Q, your pencil would move in the counterclockwise direction. Tracing $\triangle P'CQ'$ from P' to C to Q' would also result in motion in the counterclockwise direction. For this reason, we say that a rotation preserves orientation.

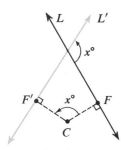

Figure 1.37

Example 1. The rotation C_x will map the line L to its image L'. To find L', rotate F, the foot of the perpendicular to L from C. Then, find the line through F' which is perpendicular to $\overline{CF'}$. This line is L', the image of L. Note that the angle between L and L' has measure x. (Try to prove this.)

Example 2. Given point A, line L, and circle C, find an equilateral triangle, AXY, such that X is in L and Y is in C.

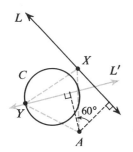

Figure 1.38

Solution. Rotate line L 60° about A to line L'. L' will intersect C in two points. Call either point Y. Point X is the preimage of Y. Since $\overline{AY}$ is the image of $\overline{AX}$, $AY = AX$. Also, $m\angle XAY = 60$. Therefore, $\triangle AXY$ is equilateral.

You will be asked to prove the following theorem in the exercises.

THEOREM 6. If lines L_1 and L_2 intersect at point C, and the angle of rotation from L_1 to L_2 is $x°$, then the result of reflecting a point in L_1 and reflecting its image in L_2 is equivalent to rotating the point $2x°$ about C. (See Figure 1.39a.) Reflecting in L_2 and then in L_1 is equivalent to the rotation C_{-2x}. (See Figure 1.39b.)

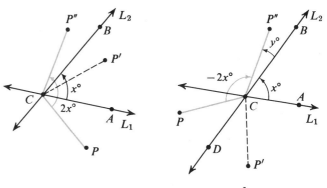

Figure 1.39   **a.**   **b.**

Figure 1.40 illustrates an important result of Theorem 6. Lines L_1, L_2, L_3, and L_4 all intersect at C. The angle from L_1 to L_2 has the same size and direction as the angle from L_3 to L_4. According to Theorem 6, successive reflections in L_1 and L_2 are equivalent to the rotation C_{2x}, and successive reflections in L_3 and L_4 are also equivalent to the rotation C_{2x}. Thus, reflecting successively in L_1 and L_2 is equivalent to reflecting successively in L_3 and L_4.

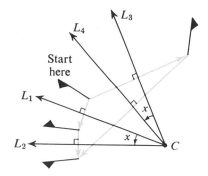

Figure 1.40

EXERCISES

1. Copy the given figure and find the image of $\triangle XYZ$ by the rotation C_{60}. Also find the image by C_{-60}.

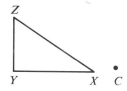

2. $ABCD$ is a square with center O. The symbol $O_{90}(A) = B$ means that B is the image of A by a 90° counterclockwise rotation about O. Find each of the following.

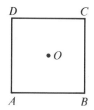

   a. $O_{-90}(A)$ **c.** $B_{-90}(A)$
   b. $O_{-270}(A)$ **d.** $D_{360}(A)$

3. $\triangle PQR$ is equilateral and has center G. Find each of the following.

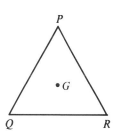

   a. $Q_{60}(R)$ **c.** $G_{120}(Q)$
   b. $Q_{-60}(P)$ **d.** $G_{360}(Q)$

4. a. Prove that if a rotation with center O maps A to A', then O lies in the perpendicular bisector of $\overline{AA'}$.

   b. Copy the figure, and use Part a to find the center of a rotation that maps A to A' and B to B'.

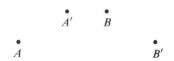

5. Triangles *ABX* and *BCY* are equilateral.

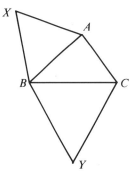

a. What symbol is used to represent a 60° clockwise rotation with center *B*?

b. Name the images of *X* and *C* by this rotation.

c. Why does $XC = AY$?

d. How large is the angle between $\overleftrightarrow{XC}$ and $\overleftrightarrow{AY}$.

6. a. Given three parallel lines, find an equilateral triangle which has one vertex in each of the lines.

b. Given three concentric circles, find an equilateral triangle which has one vertex as a point of each of the circles.

7. *ABC* and *EDC* are isosceles right triangles with right angles at *C*.

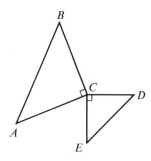

a. What symbol is used to represent a 90° counterclockwise rotation with center *C*?

b. Name the images of *B* and *E* by this rotation.

c. Why does $AD = BE$?

d. Explain why $\overleftrightarrow{AD} \perp \overleftrightarrow{BE}$.

8. The rotation C_{100} can also be described as C_{-260} and C_{460}. Give another name for each of the following rotations.

a. C_{120} **c.** C_{211} **e.** C_{0}
b. C_{-70} **d.** C_{-1000} **f.** C_{180}

9. Given point *A*, line *L* and circle *C*, as in Example 2, tell how to find a square *AXYZ* with *X* in *L* and *Z* in *C*.

10. In Figure a, a square is divided into a smaller square and four congruent right triangles. In Figure b, two right triangles are rotated about points *X* and *Y* to form Figure c. Explain how the comparison of Figures a and c illustrates the Pythagorean Theorem.

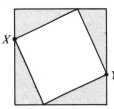

a.

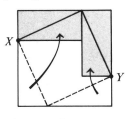

b.

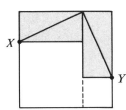

c.

11. Given two concentric circles, C_1 and C_2, and a point A which lies inside C_1 but outside C_2, construct an equilateral triangle AXY with X in C_1 and Y in C_2.

12. Prove Theorem 6 by demonstrating that for Figure 1.39a, $\measuredangle PCP'' = 2x$ and for Figure 1.39b, $\measuredangle PCP'' = -2x$. Also, show $CP = CP''$.

1.8   Translations and Glide Reflections

The shooting gallery shown in Figure 1.41 suggests the four types of isometries: reflections, rotations, translations, and glide reflections. Which targets suggest reflection? Which suggest rotation? The idea of translation or glide is suggested by the row of airplanes, or by a column of parachutists. Each airplane or parachutist glides along a geared track. The ducks do also, except they flip when hit. The relationship of the two ducks at the left and the two at the right which have been hit, suggests a glide reflection. The ducks glide along the track, and then reflect over it.

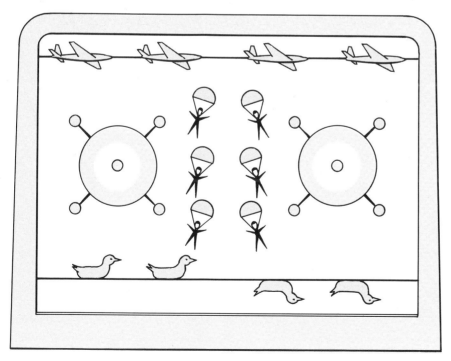

Figure 1.41

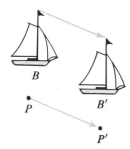

Figure 1.42

a.

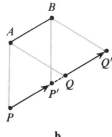

b.

Figure 1.43

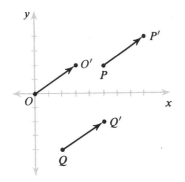

Figure 1.44

Translation is intuitively based on the notion of moving each point of the plane the same distance in the same direction. In Figure 1.42, boat B is translated to boat B'. This translation maps every point P of the plane to a unique point P' such that $\overline{PP'}$ is congruent and parallel to every segment which joins a point of B to its image in B'.

Definition. A **translation** or **glide** is a transformation such that any two points P and Q having images P' and Q' satisfy one of the following conditions.

1. $PP'Q'Q$ is a parallelogram. (See Figure 1.43a.)

2. If P, P', Q, and Q' are collinear, then there is a segment AB such that $ABP'P$ and $ABQ'Q$ are both parallelograms. (See Figure 1.43b.)

THEOREM 7. A translation is an isometry.

Proof. Assume that a translation maps P and Q to P' and Q', as shown in Figure 1.43. In Case a, $PQ = P'Q'$ because $PP'Q'Q$ is a parallelogram. In Case b, $PP' = AB = QQ'$, so that $PP' + P'Q = QQ' + P'Q$. Thus, $PQ = P'Q'$. The case when $\overline{PP'}$ and $\overline{QQ'}$ overlap can also be easily proved.

Example. A translation maps the origin to the point $(3, 2)$. Find the images of $P(5, 2)$ and $Q(2, -4)$.

Solution. To move from O to O', you must glide 3 units to the right, and 2 units upward. This same movement associates the point $P'(8, 4)$ with $P(5, 2)$. Let us check to see if P' is the image of P by the translation that maps O to O'. Both $\overline{OO'}$ and $\overline{PP'}$ have length $\sqrt{2^2 + 3^2}$, or $\sqrt{13}$. The slope of $\overline{PP'}$ is $\frac{2}{3}$, and the slope of $\overline{OO'}$ is $\frac{2}{3}$. Therefore, $\overline{OO'} \parallel \overline{PP'}$. Thus, $OO'P'P$ is a parallelogram, and $P'(8, 4)$ is the image of $P(5, 2)$ by this translation. By the same method, the image of $Q(2, -4)$ is $Q'(5, -2)$.

You will be asked to prove the following theorem in the exercises.

THEOREM 8. If $L_1 \parallel L_2$ and the distance from L_1 to L_2 is x, then the result of reflecting a point in L_1 and reflecting its image in L_2 is equivalent to translating the original point $2x$ units in the direction from L_1 to L_2.

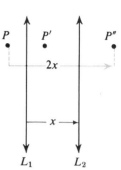

Figure 1.45

Figure 1.46 illustrates an important result of Theorem 8. Lines L_1, L_2, L_3, and L_4 are all parallel, and the distance and direction from L_1 to L_2 is the same as the distance and direction from L_3 to L_4. Theorem 8 states that successive reflections in L_1 and L_2 are equivalent to a translation of $2x$ units in the direction from L_1 to L_2. Also, by Theorem 8, successive reflections in L_3 and L_4 are equivalent to a translation of $2x$ units in the direction from L_3 to L_4. Thus, reflecting successively in L_1 and L_2 is equivalent to reflecting successively in L_3 and L_4.

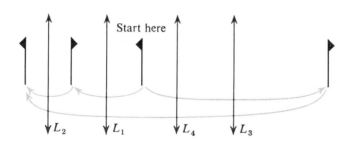

Figure 1.46

A glide reflection is exactly what the name suggests, a glide followed by a reflection. However, the line of reflection must be parallel to the direction of the glide.

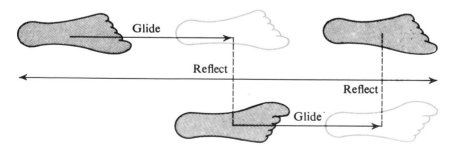

Figure 1.47

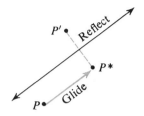

Figure 1.48

Definition. A **glide reflection** is a transformation that maps every point P of the plane to a point P' determined as follows.

1. P^* is the image of P by a glide.
2. P' is the image of P^* by a reflection in a line parallel to the direction of the glide.

Since both translations (glides) and reflections preserve distance, we shall state the following theorem.

THEOREM 9. A glide reflection is an isometry.

EXERCISES

1. A translation (or glide) maps P to P'. Find the images of A and B. Also, find the preimage of A.

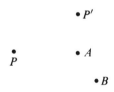

2. A translation maps the origin to the point $(5, 2)$. Where does it map the points $(1, 0)$, $(3, -7)$, $(5, 2)$, and $(-5, -2)$?

3. A translation maps the origin to the point $(4, -3)$.

    a. Where does it map the points $(-5, -1)$, $(4, -3)$, and (x, y)?

    b. What point is the preimage of the origin?

4. A translation maps $(4, 6)$ to $(7, 10)$.

    a. Where does it map $(1, 2)$ and $(-3, 0)$?

    b. What is the distance between each point and its image?

5. A glide reflection maps $\triangle ABC$ to $\triangle A'B'C'$. Locate the midpoints of $\overline{AA'}$, $\overline{BB'}$, and $\overline{CC'}$. Make a conjecture about the midpoints. Try to prove it.

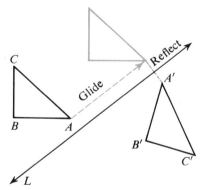

6. Copy the figure and use Exercise 5 to find the reflecting line of the glide reflection that maps $\triangle ABC$ to $\triangle A'B'C'$. Also, draw the glide image of $\triangle ABC$.

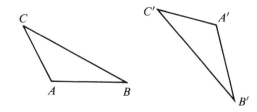

7. Describe a method of locating points X and Y, one a point of each of the circles shown, so that $\overline{XY}$ is parallel and congruent to $\overline{AB}$.

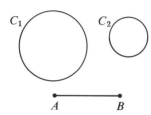

8. Describe a method of locating points X and Y, one in each of the lines shown, so that $\overline{XY}$ is parallel and congruent to $\overline{AB}$.

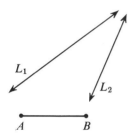

9. For the given figure, tell whether a reflection, rotation, translation or glide reflection will map Figure 1 to each of the following figures. There may be more than one possible answer.

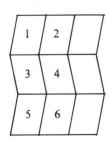

a. 2   **c.** 4   **e.** 6

b. 3   **d.** 5

10. Repeat Exercise 9 for the figure below.

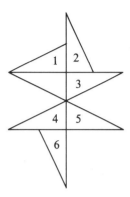

11. a. Use Figure 1.45 to prove Theorem 8.

    b. What happens to the size of the translation as the distance between L_1 and L_2 decreases? What happens when lines L_1 and L_2 coincide and the distance between them is zero? (This translation is called the *zero translation or identity*.)

12. Use Theorem 8 to explain how a periscope works.

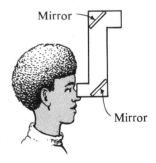

13. Study the airplanes and ducks of the shooting gallery in Figure 1.41. Is orientation preserved or reversed by a glide? By a glide reflection?

14. Study Figure 1.47. What single transformation is equivalent to the result of using a glide reflection twice?

15. The following are some examples of the graphic art of M. C. Escher (1898–1972). Explain where the ideas of reflection, rotation, translation and glide reflections are used.

a.

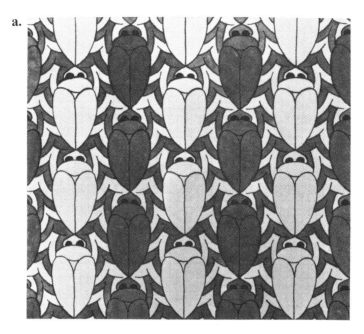

b.

1.9   Symmetry

Each of the figures shown below has symmetry.

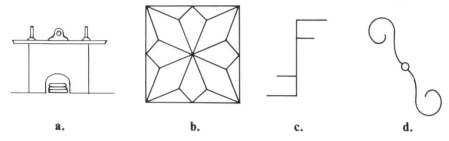

a.   b.   c.   d.

Figure 1.49

If a vertical line were drawn through the middle of Figure 1.49a, the left half and the right half would be mirror images of each other. This figure is said to have line symmetry. Figure 1.49b also is said to have line symmetry. In fact, this figure has a total of four lines of symmetry. If the figure is reflected in any of these lines, its image is the same as the original figure. The symmetry of Figures 1.49c and d is called point symmetry, because a half-turn about the middle point of either figure will result in an image which is the same as the original figure.

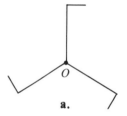

a.

Definition. A figure F has **line symmetry** if and only if there is a line L such that $\mathsf{R}_L(F) = F$. Line L is called a *line of symmetry*.

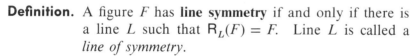

Definition. A figure F has **point symmetry** if and only if there is a point O such that $\mathsf{H}_O(F) = F$. Point O is called a *point of symmetry*.

b.

In addition to line and point symmetry, there is a third type of symmetry known as rotational symmetry. This type of symmetry is illustrated in Figure 1.50.

For Figure 1.50a, rotations of 120° and 240° about its center point, O, will map the figure onto itself. For Figure 1.50b, rotations of 72°, 144°, 216°, and 288° will map the figure onto itself. Similarly, rotations of 90°, 180°, and 270° about the center of Figure 1.50c map it to itself. Of course, a 180° rotation is the same as a half-turn, so Figure 1.50c also has point symmetry.

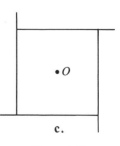

c.

Figure 1.50

Definition. A figure has **rotational symmetry** if and only if there is a rotation, which is not equivalent to the zero rotation, that maps the figure onto itself.

Definition. A **symmetry** of a figure is any isometry that maps the figure onto itself.

Note that this definition of symmetry allows us to consider the identity mapping as a symmetry. The identity mapping associates every point with itself. We shall represent this mapping with the symbol I.

Using the definition of symmetry, we can determine that a rectangle has four symmetries. (See Figure 1.51.) Two are line reflections, the third is a point reflection (half-turn), and the fourth is the identity, I. While it may seem a bit generous to call the identity mapping a symmetry, we shall see, later, the reason for doing so. Thus, a rectangle has four symmetries.

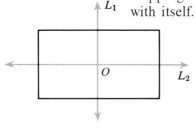

Symmetries of a rectangle:

R_1, R_2, H_O, I

Figure 1.51

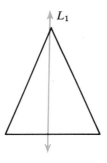

Symmetries of an
isosceles triangle:

R_1, I

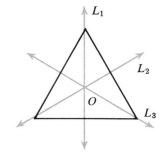

Symmetries of an
equilateral triangle:

$R_1, R_2, R_3, O_{120}, O_{240}, I$

Figure 1.52

Since the identity mapping is a symmetry of every figure, a figure must have at least one symmetry. An isosceles triangle has a second symmetry. This symmetry is a reflection in the line, L_1, which is the perpendicular bisector of the base of the triangle. (See Figure 1.52.) Thus, an isosceles triangle has two symmetries. An equilateral triangle has six symmetries. Three are reflections in the perpendicular bisectors of the sides of the triangle. (See Figure 1.52.) Two are rotational symmetries, and the sixth is the identity mapping.

EXERCISES

1. Which of the following figures have point symmetry, which have line symmetry, and which have rotational symmetry?

a. H   b. X   c. ⬭   d. N   e. $ f. ⬡

2. The graphs of some familiar equations are shown below. Which graphs have line symmetry and which have point symmetry? Give the equations of all lines of symmetry, and the coordinates of all points of symmetry. Compare graphs a and b, d and e, and g and h. Describe the effect up and down translations have on the equation of the graph. Similarly, compare graphs a and c, d and f, and g and i. Describe the effect the reflection in the line $y = x$ has on the equation of a graph.

a.

$y = x^2$

d.
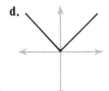

$y = |x|$

g.
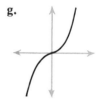

$y = x^3$

b.
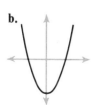

$y = x^2 - 4$

e.
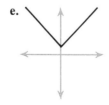

$y = |x| + 1$

h.
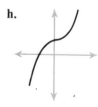

$y = x^3 + 2$

c.

$x = y^2$

f.
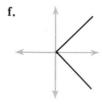

$x = |y|$

i.
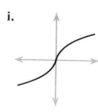

$x = y^3$

3. Print the alphabet as capital letters. Which letters have point symmetry, and which have line symmetry?

4. The following illustrate symmetry in nature. Discuss each, and determine how many symmetries each has.

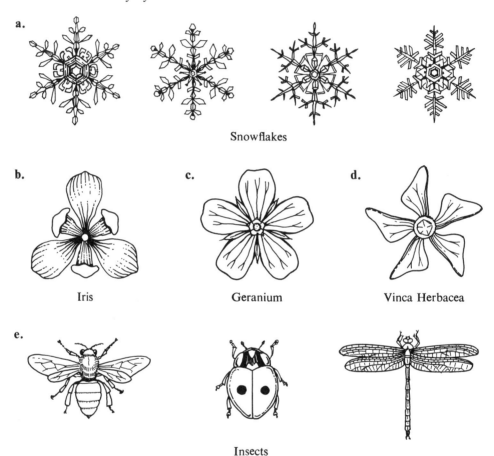

a.

Snowflakes

b.

Iris

c.

Geranium

d.

Vinca Herbacea

e.

Insects

5. **a.** Most naturalists refer to line symmetry as bilateral symmetry. Why does this seem an appropriate name?

   b. Which of the following have bilateral symmetry: a maple leaf, a moose, a jelly fish, an octopus, a mosquito, a daddy longlegs, a human being?

   c. Which of the items listed in part b have symmetries other than line symmetry?

6. Describe the symmetries of a parallelogram, a rhombus, and a square. Recall that a square is also a parallelogram, a rhombus, and a rectangle. Thus, the symmetries of the square should include the symmetries of the other three.

7. Describe the symmetries of each of the following figures.

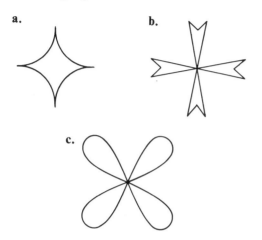

a.

b.

c.

8. The hexagon *ABCDEF* with center *O* has the point symmetry H_O.

a. What is the image of $\overline{AB}$? Explain why opposite sides of the hexagon have the same length.

b. What is the image of $\angle ABE$? Explain why opposite sides of the hexagon are parallel.

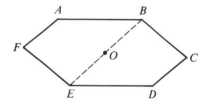

1.10 The Fundamental Theorems of Isometries

THEOREM 10. (The First Fundamental Theorem of Isometries) An isometry of the plane is determined by three noncollinear points and their images.

This theorem is typical of many statements in mathematics. Here are some similar examples.

> A line is determined by two points.
> A plane is determined by three noncollinear points.
> A circle is determined by three noncollinear points.

There are two things implied by Theorem 10. First, given any three noncollinear points, *A*, *B*, and *C*, and their images *A'*, *B'*, and *C'*, we should be able to find the image of every point in the plane. Second, there is exactly one isometry that maps *A*, *B*, and *C*, to *A'*, *B'*, and *C'*.

Let us now demonstrate how to find the image of any point, *P*, given that the isometry maps the noncollinear points *A*, *B*, and *C* to *A'*, *B'*, and *C'*. Since the isometry maps *A* to *A'*, it must map *P* to some point

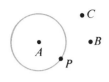

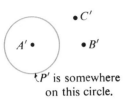

P' is somewhere
on this circle.

a.

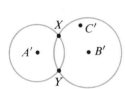

b.

Figure 1.53

P' of the circle with center *A'*, and radius *AP*. (See Figure 1.53a.)

Since the isometry also maps *B* to *B'*, then the image of *P* must also be a point of the circle with center *B'* and radius *BP*. These circles intersect in two points, *X* and *Y*, so either point could be the image of *P*. (See Figure 1.53b.)

To determine which of the points, *X* or *Y*, is the image of *P*, observe whether *P* and *C* are on the same side, or opposite sides of $\overleftrightarrow{AB}$. In this case, *P* and *C* are on opposite sides of $\overleftrightarrow{AB}$. Therefore, *P'* and *C'* must be on opposite sides of $\overleftrightarrow{A'B'}$, so *P'* must be point *Y*.

Now, we shall show that there is exactly one isometry which maps *A*, *B*, and *C* to *A'*, *B'*, and *C'*. Given the three noncollinear points and their images, we have shown that we can find the image of any point in the plane. Therefore, at least one such isometry exists. Assume that there are, in fact, two different isometries, F_1 and F_2, which map *A*, *B*, and *C* to *A'*, *B'*, and *C'*. Since F_1 and F_2 are different mappings, there must be one point *P* for which $F_1(P) \neq F_2(P)$. Let $F_1(P) = P_1$ and $F_2(P) = P_2$. Consider the following argument.

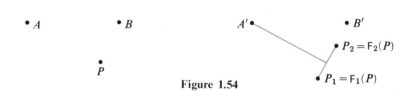

Figure 1.54

1. $AP = A'P_1$, because F_1 is an isometry.
2. $AP = A'P_2$, because F_2 is an isometry.
3. Thus, $A'P_1 = A'P_2$ and *A'* is in the perpendicular bisector of $\overline{P_1P_2}$.
4. Similarly, *B'* and *C'* are in the perpendicular bisector of $\overline{P_1P_2}$.
5. *A'*, *B'*, and *C'* are all points of the perpendicular bisector of $\overline{P_1P_2}$. Therefore, *A'*, *B'*, and *C'* are collinear.

Since F_1 and F_2 are isometries, and isometries preserve collinearity, neither can map the noncollinear points *A*, *B*, and *C* to the collinear points *A'*, *B'*, and *C'*. Therefore, our assumption is false, and there is only one isometry which maps *A*, *B*, and *C* to *A'*, *B'*, and *C'*.

THEOREM 11. (The Second Fundamental Theorem of Isometries) Every isometry is equivalent to a succession of at most three line reflections.

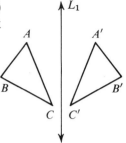

L_1

Proof. By Theorem 10, an isometry F is determined by $\triangle ABC$ and $\triangle A'B'C'$. It is possible that one line reflection will map $\triangle ABC$ to $\triangle A'B'C'$, as illustrated in Figure 1.55.

If the single reflection in L_1 does not map $\triangle ABC$ to $\triangle A'B'C'$, then let $R_1(B) = B_1$ and $R_1(C) = C_1$. Follow R_1 by a reflection in L_2, the perpendicular bisector of $\overline{B_1B'}$, as shown in Figure 1.56a. R_2 maps B_1 to B'. By the first reflection, $AB = A'B_1$. Since we are given $AB = A'B'$, then $A'B_1 = A'B'$. Therefore, A' is on the perpendicular bisector of $\overline{B_1B'}$, and $R_2(A') = A'$. If the original triangles, $\triangle ABC$ and $\triangle A'B'C'$, have the same orientation, $R_2(C_1) = C'$. If they have opposite orientation, another reflection is needed.

Figure 1.55

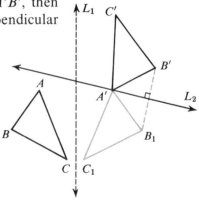

Let $R_2(C_1) = C_2$ and let L_3 be the perpendicular bisector of $\overline{C'C_2}$. By R_3, the reflection in L_3, C_2 is mapped to C'. To show R_3 keeps A' and B' fixed, we use R_1 and R_2 and the given isometry to write $A'C_2 = A'C_1 = AC = A'C'$ and $B'C_2 = B_1C_1 = BC = B'C'$. Thus, A' and B' are equidistant from C_2 and C' and so are contained in L_3. Therefore, R_3 keeps A' and B' fixed and maps $\triangle A'B'C_2$ to $\triangle A'B'C'$.

a.

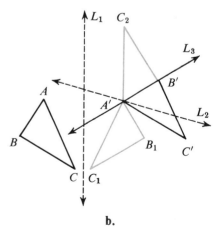

b.

Figure 1.56

Three reflections, at most, are needed to map $\triangle ABC$ to $\triangle A'B'C'$, where $\triangle A'B'C'$ is the image of $\triangle ABC$ by the isometry F. By the First Fundamental Theorem of Isometries, there is only one isometry F which maps $\triangle ABC$ to $\triangle A'B'C'$. Thus, the isometry F is equivalent to the succession of reflections in the sense that every point will be mapped to the same image point by either the isometry or the succession of reflections.

THEOREM 12. (The Third Fundamental Theorem of Isometries) Every isometry of the plane is either a line reflection, a rotation, a translation, or a glide reflection.

This theorem states that a fifth type of isometry cannot exist, that we have studied all possible types. We may describe some isometry in different words, but its effect will be the same as some reflection, rotation, translation, or glide reflection. This theorem is based on the fact that an isometry F is a succession of at most three reflections (Theorem 11).

1. If there is but one line reflection, then F is a reflection.
2. If there are two reflections, and the lines of reflections are parallel, then F is a translation. (Theorem 8)
3. If there are two reflections, and the lines of reflection are not parallel, then F is a rotation. (Theorem 6)
4. If there are three reflections and all three lines of reflection are parallel, then F is a reflection.
5. If there are three reflections and the three lines of reflection are concurrent, then F is a reflection.
6. If there are three reflections and the lines of reflection are not all parallel or all concurrent, then F is a glide reflection.

Proofs of Cases 4, 5, and 6 are considered in Exercises 17, 18, and 19 of Section 2.1.

EXERCISES

1. In each figure, a reflection maps A to A'. Copy each figure, locate the line of reflection and P'. Is a reflection determined by a point and its image?

a.
                • A'

  • A

            • P

b.
  • $A = A'$

         • P

2. A half-turn maps A to A'. Copy the figure and locate P'. Is a half-turn determined by a point and its image?

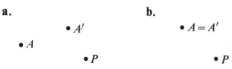

3. An isometry maps $\triangle ABC$ to $\triangle A'B'C'$. Copy the figure, and locate the images of P and Q using only a compass.

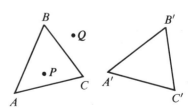

4. How many point and image pairs are needed to determine a translation? How many determine a rotation?

2

The Algebra of Transformations

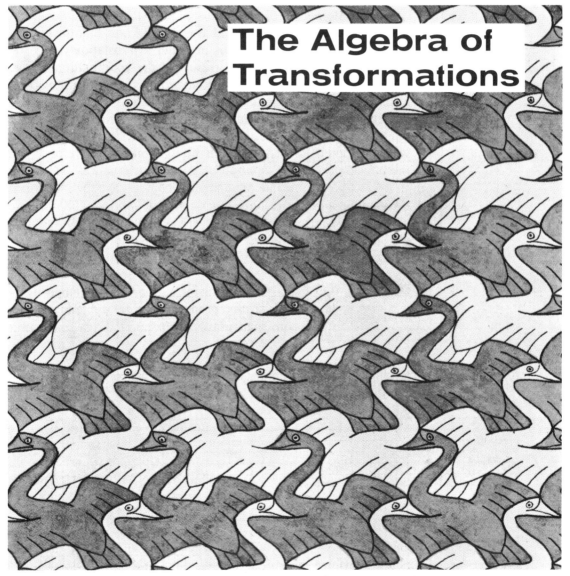

2.1   The Composite (Product) of Mappings

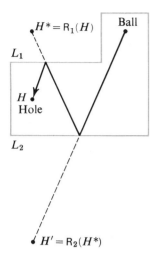

Figure 2.1

Recall the problem of how to hit a hole-in-one on a miniature golf course. For the hole shown in Figure 2.1, the solution uses successive reflections in parallel lines. By Theorem 8, we know that this is equivalent to a translation. Thus, if the golfer mentally translates the hole, H, to H', and aims for H', he should get a hole-in-one.

The combination of reflections used in this problem suggests the topic of discussion for this chapter. We know that the combination of two reflections results in either a translation or a rotation. But, what would be the result of the combination of two translations, or two rotations? What would be the result of combining two different kinds of isometries, such as a translation and a half-turn? As we investigate the interrelationships of isometries, our point of view will become more algebraic. In Chapter 1, our view of isometries was predominantly geometric, as we observed the effects of a transformation on a geometric figure. Now, our terminology will resemble that of an algebra course. We shall use terms such as *commutative, associative, identity,* and *inverses.* Before we define any of these terms, we shall state what we mean by a combination or succession of transformations, called a composite of transformations, or a product of transformations.

> **Definition.** Let F and G be transformations. The **composite** of G and F, or the **product** of G and F, is a transformation that maps each point P to the point $P' = G(F(P))$.

Therefore, in order to find $G(F(P))$, we must first find $P^* = F(P)$, and then find $P' = G(P^*)$. P' is the image of P by the composite of G and F. It is customary to denote this composite as GF. Thus, GF(P) really means $G(F(P))$. The operation of forming composites is called composition.

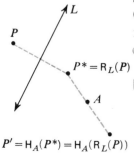

Figure 2.2

Example 1. The composite (or product) of the half-turn H_A and the reflection R_L is the transformation $H_A R_L$. It will map each point P to $H_A(R_L(P))$. (See Figure 2.2.)

Note that although we would read $H_A R_L$ from left to right, the order in which we perform the transformations is right to left. The half-turn follows the reflection. To find the image of any point by this composite transformation, first R_L is applied and then H_A. Similarly, for the composite of G and F, G follows F. For this reason, the symbol GF is sometimes read as "G follows F."

Example 2. Figure 2.3a shows the image of P by the transformation R_2R_1. Figure 2.3b shows the image of P by the transformation R_1R_2. The figures indicate that the order in which transformations are combined is important. In this example, $R_2R_1 \neq R_1R_2$.

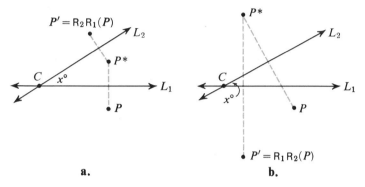

a.　　　　　　　　　　　**b.**

Figure 2.3

Actually, the fact that $R_2R_1 \neq R_1R_2$ should not be surprising. From Theorem 6 we know that a composite of reflections in intersecting lines is a rotation through twice the angle from the first line of reflection to the second. In this case, we would write the following.

$$R_2R_1 = C_{2x} \text{ and } R_1R_2 = C_{-2x}$$

Example 2 illustrates that the composition of transformations is not a commutative operation.

> **Definition.** An operation $*$ that combines elements of a set S is a **commutative** operation if and only if $a * b = b * a$ for all a and b in S.

Example 3. **a.** $a + b = b + a$ for every a and b in R, the set of real numbers. Therefore, addition of real numbers is commutative.

b. $a \cdot b = b \cdot a$ for every a and b in R. Therefore, multiplication of real numbers is commutative.

c. $a - b \neq b - a$ for every a and b in R. Therefore, subtraction of real numbers is not commutative.

d. FG $\neq$ GF for every pair of transformations, F and G. Therefore, composition of transformations is not commutative.

Parts b and d of Example 3 indicate that the composition of transformations is unlike the multiplication of real numbers in one way. However, there are many ways in which the operations are alike. Both composition and multiplication are associative operations, and both have identity elements. For this reason, composition is sometimes thought of as multiplication, and the composite of two transformations may be called the product of the transformations.

> **Definition.** An operation $*$ that combines elements of a set S is **associative** if and only if $(a * b) * c = a * (b * c)$ for all a, b, and c in S.

Example 4. **a.** $(ab)c = a(bc)$ for every a, b, and c in R, so multiplication of real numbers is associative.

b. $(a + b) + c = a + (b + c)$ for every a, b, and c in R, so addition of real numbers is associative.

c. $(A \cup B) \cup C = A \cup (B \cup C)$ for all sets, so the operation of forming the union of sets is associative.

d. $(a - b) - c \neq a - (b - c)$ for every a, b, and c in R, so subtraction of real numbers is not associative.

We shall now demonstrate that the composition of transformations is associative. Consider an arbitrary point, P. Let $H(P) = P'$, $G(P') = P''$ and $F(P'') = P'''$. We want to show that $(FG)H = F(GH)$. We shall do this by showing that each transformation has the same effect on P.

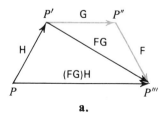
a.

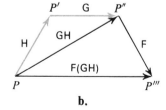
b.

Figure 2.4

Since $\mathsf{H}(P) = P'$, and $\mathsf{FG}(P') = P'''$, then $(\mathsf{FG})\mathsf{H}$ maps P to P'''. (See Figure 2.4a.) Since $\mathsf{GH}(P) = P''$, and $\mathsf{F}(P'') = P'''$, then $\mathsf{F}(\mathsf{GH})$ maps P to P'''. Both $(\mathsf{FG})\mathsf{H}$ and $\mathsf{F}(\mathsf{GH})$ have the same effect on an arbitrary point P; therefore we may conclude that they are the same transformation. We shall state this result as a theorem.

THEOREM 13. The composition of transformations is associative.

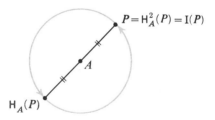

Figure 2.5

When a transformation F is combined with itself, the product, FF, is usually written F^2. Thus, if T is the translation which maps A to B, then T^2 maps A to C, and T^3 maps A to D, as shown in Figure 2.5.

When a line reflection or a point reflection is combined with itself, the resulting transformation leaves all points fixed, as shown in Figure 2.6. This transformation is the identity transformation, I, and $\mathsf{I}(P) = P$ for every point P. The combination of the identity with any transformation F accomplishes no more than F does by itself. That is, $\mathsf{FI}(P) = \mathsf{F}(\mathsf{I}(P)) = \mathsf{F}(P)$, so $\mathsf{FI} = \mathsf{F}$.

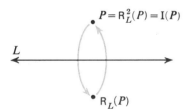

Figure 2.6

Definition. Let $*$ be an operation that combines elements of a set S. The element I of S is the **identity element** for the operation $*$ if and only if $a * I = a = I * a$, for every element a in S.

Example 5. **a.** $a + 0 = a = 0 + a$ for every a in R, so 0 is the identity element for addition of real numbers.

b. $a \cdot 1 = a = 1 \cdot a$ for every a in R, so 1 is the identity element for multiplication of real numbers.

c. $\mathsf{AI} = \mathsf{A} = \mathsf{IA}$ for every transformation A, so I is the identity element for composition of transformations.

d. $a - 0 = a$, but $0 - a \neq a$ for every a in R, so 0 is not the identity element for subtraction of real numbers.

Once it is determined that a set has an identity element, it is natural to give attention to pairs of elements that produce the identity.

> **Definition.** Let $*$ be an operation that combines elements of a set S, which has the identity element I. The elements a and b of S are **inverses** if and only if $a * b = I = b * a$.

Example 6. **a.** $2 + (-2) = 0$, the identity element for addition, so 2 and -2 are additive inverses.

b. $2 \times \dfrac{1}{2} = 1$, the identity element for multiplication, so 2 and $\dfrac{1}{2}$ are multiplicative inverses.

c. $C_{90}C_{-90} = C_0 = I$ and $C_{-90}C_{90} = C_0 = I$, where I is the identity transformation, so C_{90} and C_{-90} are inverse transformations.

d. $R_L R_L = I$, the identity transformation, so R_L is its own inverse.

The inverse of a transformation M is usually denoted M^{-1}. Thus, $MM^{-1} = I = M^{-1}M$. An analogous statement from algebra may already be familiar to you. Often the multiplicative inverse of a number, m, is written as m^{-1}. Thus, $mm^{-1} = 1 = m^{-1}m$.

Since every transformation is a one-to-one mapping, every point of the plane has exactly one preimage point. Because of this, every transformation has an inverse. Figure 2.7 illustrates two examples.

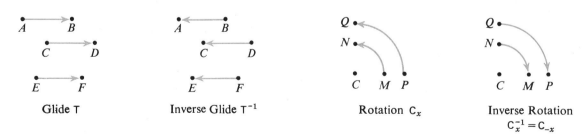

| Glide T | Inverse Glide T^{-1} | Rotation C_x | Inverse Rotation $C_x^{-1} = C_{-x}$ |

Figure 2.7

Figure 2.8a shows that the inverse of putting on your socks and then putting on your shoes is taking off your shoes and then taking off your socks. This suggests that the inverse of FG is $G^{-1}F^{-1}$, or that $(FG)^{-1} = G^{-1}F^{-1}$.

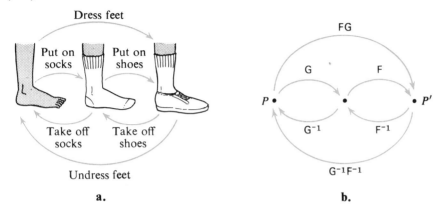

Figure 2.8

THEOREM 14. If F and G are any two transformations, then $G^{-1}F^{-1}$ is the inverse of FG.

  Proof. We can prove this theorem by showing that the product of FG and $G^{-1}F^{-1}$ equals I.

$$(FG)(G^{-1}F^{-1}) = F(GG^{-1})F^{-1} \quad \text{(Theorem 13)}$$
$$= FIF^{-1}$$
$$= FF^{-1}$$
$$= I$$

  Therefore, FG and $G^{-1}F^{-1}$ are inverses.

EXERCISES

1. In forming the composite FG, which transformation is applied first, F or G?

2. Copy the given figure, and locate the image of P by R_1R_2 and by R_2R_1.

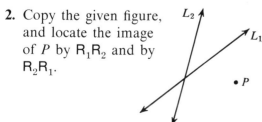

3. If X is a person, let $F(X)$ be the father of X, and let $M(X)$ be the mother of X. Match each composite in the first column with the appropriate name in the second column.

FM	paternal grandfather
FF	maternal grandmother
MF	paternal grandmother
MM	maternal grandfather

4. Let R_x and R_y be reflections in the x- and y-axes. Show that the image of any point $P(x, y)$ is the same under the transformation $R_x R_y$ as it is under $R_y R_x$. What single transformation is equivalent to $R_x R_y$ and $R_y R_x$?

5. **a.** If $L_1 \parallel L_2$, does $R_1 R_2 = R_2 R_1$?

   b. If $L_1 \perp L_2$, does $R_1 R_2 = R_2 R_1$?

   c. Is the composition of reflections commutative?

6. Copy the given figure, and locate the image of P under each of the following transformations.

   a. $H_A R_L$ **b.** $R_L H_A$ **c.** $H_A{}^2$ **d.** $R_L{}^2$

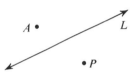

7. Copy the given figure and locate the image of P under the transformations $H_B H_A$ and $H_A H_B$.

   $\bullet^A$ $B_\bullet$

   $\bullet P$

8. Sketch the triangle with vertices $A(0, 0)$, $B(3, 6)$ and $C(5, 2)$. Find the images of $\triangle ABC$ under the transformations FG and GF, where G:$(x, y) \rightarrow (2x, 2y)$ and F:$(x, y) \rightarrow (x - 3, y + 1)$.

9. Which of the following products of transformations are the same as the identity mapping?

   a. C_{360} **d.** $R_1 R_2 R_2 R_1$

   b. C_{-720} **e.** $C_{120} C_{240}$

   c. $R_1{}^2$

10. In each of the following, $a * b$ is the result of combining two positive integers, a and b, by some operation, $*$. In each case, tell if $*$ is commutative.

    a. $a * b =$ average of a and b

    b. $a * b = a^b$

    c. $a * b =$ greatest common divisor of a and b

    d. $a * b = 2a + b$

    e. $a * b = (a - b)^2$

    f. $a * b = (a - b)^3$

    g. $a * b = |a - b|$

11. Which of the operations listed in Exercise 10 do you think are associative? (Hint. Let a, b, and c equal specific numbers and check whether or not $(a * b) * c = a * (b * c)$. If not, this shows that the operation is not associative. On the other hand, if the equation is true for a few examples, it will *probably* hold for all examples, but you are not asked to prove this.)

12. State the additive and multiplicative inverses of each of the following.

    a. 5   **c.** 1   **e.** $\dfrac{2}{5}$

    b. -7 **d.** 0   **f.** $-\dfrac{1}{2}$

13. Name the transformation which is the inverse of each of the following.

    a. C_{135} **c.** $R_1 R_2$ **e.** R_L

    b. H_A **d.** I   **f.** FG

14. Show that $H_A H_B$ and $H_B H_A$ are inverses by the following two methods.

    a. Show $(H_A H_B)(H_B H_A) = I$ by using the associative property of composition.

    b. Use Theorem 14.

15. Consider the projection of the plane on line L, as shown.

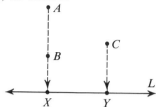

a. Name the images of A and B.

b. Explain why this mapping has no inverse.

16. Classify each of the following transformations as a half-turn, reflection, translation, or rotation. Give the inverse in each case.

a. $F:(x, y) \rightarrow (-x, y)$

b. $F:(x, y) \rightarrow (-x, -y)$

c. $F:(x, y) \rightarrow (-y, x)$

d. $F:(x, y) \rightarrow (x, y + 3)$

e. $F:(x, y) \rightarrow (x + 10, y)$

f. $F:(x, y) \rightarrow (x + 10, y + 3)$

Exercises 17–19 are proofs for Cases 4, 5, and 6 of Theorem 12. Supply the missing reasons, or complete the proof as requested.

17. Given that $L_1 \parallel L_2 \parallel L_3$, prove $R_3R_2R_1$ is a reflection.

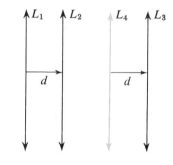

Proof. 1. Let L_4 be the line parallel to L_3 such that the distance and direction from L_4 to L_3 is the same as the distance and direction from L_1 to L_2.

2. Then, $R_3R_4 = R_2R_1$. (This is a result of Theorem 8.)

3. Therefore, $R_3R_2R_1 = R_3(R_2R_1)$ Why?
$= R_3(R_3R_4)$ Why?
$= (R_3R_3)R_4$ Why?
$= IR_4$ Why?
$= R_4$ Why?

18. Given that L_1, L_2, and L_3 are concurrent at O, prove $R_3R_2R_1$ is a reflection.

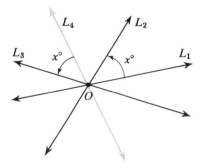

Proof. 1. Let L_4 be the line through O such that the angle from L_4 to L_3 is the same size and direction as the angle from L_1 to L_2.

2. Then, $R_3R_4 = R_2R_1$. (This is a result of Theorem 6.)

3. Complete this proof. It is similar to step 3 of Exercise 17.

19. Given that L_1, L_2, and L_3 are not all parallel or concurrent, prove that $R_3R_2R_1$ is a glide reflection.

Proof. 1. A is the point of intersection of L_1 and L_2. Let L_4 be the line containing the perpendicular from A to L_3.

2. Let L_5 be the line through A such that the angle from L_5 to L_4 is the same size and direction as the angle from L_1 to L_2. Thus, $R_2R_1 = R_4R_5$.

3. B is the point of intersection of L_3 and L_4. Let L_6 and L_7 be the lines through B which are parallel and perpendicular to L_5, respectively. Thus, R_6R_5 is a glide, and $R_7(R_6R_5)$ is a glide reflection. Why?

4. $R_3R_2R_1 = R_3(R_2R_1)$ Why?
$= R_3(R_4R_5)$ Why?
$= (R_3R_4)R_5$ Why?
$= (R_7R_6)R_5$ Why?
$= R_7(R_6R_5)$ Why?

Therefore, $R_3R_2R_1$ is a glide reflection.

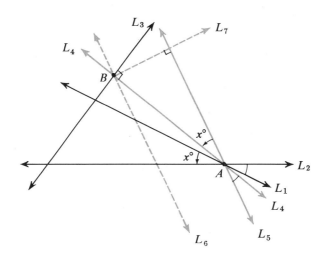

20. In a coordinate plane, let L_1 be the line with equation $y = 1$, and let L_2 be the line with equation $y = -4$. Find the images of each of the following points by the transformation R_2R_1.

  a. $A(2, 1)$ **b.** $B(-1, -1)$ **c.** $C(-6, 0)$ **d.** $D(x, y)$

21. Find the images of each of the points listed in Exercise 20 by the transformation R_1R_2.

2.2   The Algebra of Translations

A translation that maps point A to point B may be represented in a figure by an arrow from A to B. (See Figure 2.9.)   The symbol $\overrightarrow{AB}$ is used to represent either the arrow from A to B, or the translation that maps A to B. If this translation also maps P to Q, then both $\overrightarrow{PQ}$ and $\overrightarrow{AB}$ represent the same translation.   In this case, we shall write $\overrightarrow{PQ} = \overrightarrow{AB}$, although we do not mean that the two arrows are the same.   Rather, it is like saying that $\frac{4}{6} = \frac{2}{3}$.   Both

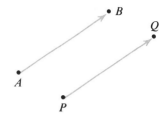

Figure 2.9

$\frac{4}{6}$ and $\frac{2}{3}$ represent the same number, just as $\overrightarrow{AB}$ and $\overrightarrow{PQ}$ represent the same translation.   Thus, $\overrightarrow{PQ} = \overrightarrow{AB}$ when the arrow from P to Q has the same length and direction as the arrow from A to B.   Often these arrows are called vectors. $\overrightarrow{AB}$ may be read as "arrow AB" or as "vector AB."

Suppose you are playing shuffleboard on the deck of a ship. (See Figure 2.10.)   If the ship is not moving, the glide of the puck from A to B describes a translation, T.   Now suppose that the ship is moving toward the east.   As the puck travels from A to B, the ship moves eastward, so that the shuffleboard court moves from $\overline{AB}$ to $\overline{CD}$.   (See Figure 2.11.) The ship's movement during this time describes another translation, S.   The composite of the translations S and T will be still another translation, V, that glides the puck from A to D.

The translation T, which sends the puck from A to B is represented in Figure 2.11 by an arrow from A to B.   We may use the symbol $\overrightarrow{AB}$ to refer either to this arrow or to the translation T.   Since T maps C to D, T may also be represented by the symbol $\overrightarrow{CD}$.   Similarly, $\overrightarrow{AC}$ and $\overrightarrow{BD}$ refer to the translation S.   In considering the composite of S and T, it makes no difference whether we think of the ship's glide east following the puck's glide north (ST = V) or the puck's glide north following the ship's glide east (TS = V).   Later, we shall prove that the order in which translations are combined does not matter.

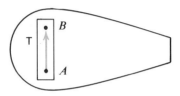

Figure 2.10

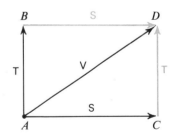

Figure 2.11

Example 1. An airplane is traveling east with a velocity of 300 miles per hour. Its motion from point X to point Y in an hour describes a translation, T, which can be represented by $\overrightarrow{XY}$. Likewise, in an hour, a wind blowing from the southwest at 50 miles per hour describes a translation, S, represented by $\overrightarrow{XZ}$ or $\overrightarrow{YW}$. The combined effect of these two translations is a new translation, V, that moves the airplane from X to W in one hour.

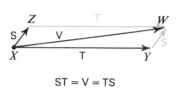

ST = V = TS

Figure 2.12

We can express the result of Example 1 by this equation.

$$\overrightarrow{XY} + \overrightarrow{YW} = \overrightarrow{XW}$$

This is a short way of saying that the two translations, which map X to Y and Y to W, combine to map X to W. It should be understood that the plus sign in this equation does not refer to standard addition. Rather, the plus sign represents a method by which we can add arrows that represent translations. The result of Figure 2.11 can be written as $\overrightarrow{AB} + \overrightarrow{AC} = \overrightarrow{AD}$.

Example 2. In Figure 2.13a, the translation mapping P to Q combined with the translation mapping Q to R, is equivalent to the translation mapping P to R. Therefore, we may write the equation $\overrightarrow{PQ} + \overrightarrow{QR} = \overrightarrow{PR}$. Similarly, for Figure 2.13b, $\overrightarrow{JK} + \overrightarrow{KM} = \overrightarrow{JM}$.

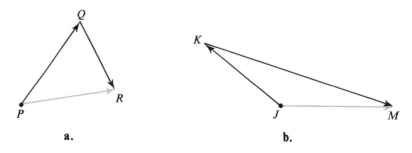

a.　　　　　　　　　　**b.**

Figure 2.13

In Examples 1 and 2 we have illustrated two methods of adding arrows. Figure 2.14a shows the parallelogram method, which applies when the arrows are situated end to end. The triangle method shown in Figure 2.14b, applies when the arrows are situated tip to end.

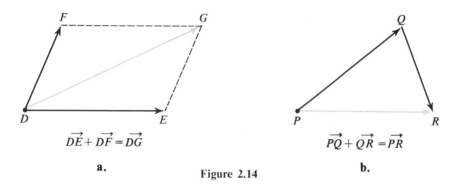

$$\overrightarrow{DE} + \overrightarrow{DF} = \overrightarrow{DG} \qquad\qquad \overrightarrow{PQ} + \overrightarrow{QR} = \overrightarrow{PR}$$

a.   **Figure 2.14**   **b.**

Three or more translations can be added by combining two at a time. In Figure 2.15, to find $\overrightarrow{AB} + \overrightarrow{BC} + \overrightarrow{CD}$, first find $\overrightarrow{AB} + \overrightarrow{BC} = \overrightarrow{AC}$. Then, $\overrightarrow{AB} + \overrightarrow{BC} + \overrightarrow{CD} = (\overrightarrow{AB} + \overrightarrow{BC}) + \overrightarrow{CD} = \overrightarrow{AC} + \overrightarrow{CD} = \overrightarrow{AD}$.

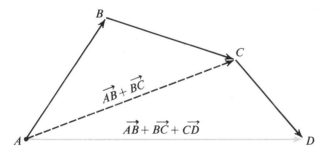

Figure 2.15

If a translation T is represented by $\overrightarrow{AB}$, then the translation T^2 is represented by $\overrightarrow{AB} + \overrightarrow{AB}$. We shall write this more simply as $2\overrightarrow{AB}$. Similarly, T^3 is represented by $3\overrightarrow{AB}$, an arrow in the same direction as $\overrightarrow{AB}$, but three times as long. Finally, if T maps A to B, then T^{-1} maps B to A and can be represented by $-\overrightarrow{AB}$ or $\overrightarrow{BA}$.

In Figure 2.16, $AB = BC = CD$. Therefore, T may be represented by $\overrightarrow{AB}, \overrightarrow{BC},$ or $\overrightarrow{CD}$. T^2 may be represented by $\overrightarrow{AC}$ or by $\overrightarrow{BD}$. T^3 may be represented by $\overrightarrow{AD}$, and T^{-1} may be represented by $\overrightarrow{BA}, \overrightarrow{CB},$ or $\overrightarrow{DC}$.

$$A \quad B \quad C \quad D$$

$$\mathsf{T} \;\;= \overrightarrow{AB}$$
$$\mathsf{T}^2 \;= 2\overrightarrow{AB} = \overrightarrow{AC}$$
$$\mathsf{T}^3 \;= 3\overrightarrow{AB} = \overrightarrow{AD}$$
$$\mathsf{T}^{-1} = -\overrightarrow{AB} = \overrightarrow{BA}$$

Figure 2.16

Now, we shall state two theorems whose proofs are left as exercises.

THEOREM 15. The composite of two translations is a translation.

THEOREM 16. The composition of translations is commutative.

Theorem 15 implies that the glides of the ship and the puck, discussed earlier in this section, really do combine to produce another glide. Theorem 16 implies that the order in which these glides are combined does not matter.

EXERCISES

1. In which of the following cases do the pair of arrows appear to represent the same translation?

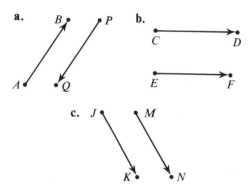

2. *ABCD* is a parallelogram. Using the properties of a parallelogram, name an arrow equivalent to each of these.

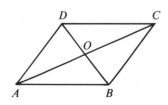

a. $\overrightarrow{AB}$ c. $\overrightarrow{OD}$ e. $\overrightarrow{AD} + \overrightarrow{AB}$

b. $\overrightarrow{BC}$ d. $\overrightarrow{OA}$ f. $\overrightarrow{AD} + \overrightarrow{DC}$

3. *ABCD* and *ADEF* are both parallelograms. Name *all* arrows equivalent to each of the following.

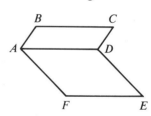

a. $\overrightarrow{BC}$ d. $\overrightarrow{AD} + \overrightarrow{AF}$

b. $\overrightarrow{AB}$ e. $\overrightarrow{FE} + \overrightarrow{FB}$

c. $\overrightarrow{BF}$ f. $\overrightarrow{AF} + \overrightarrow{FE}$

4. Copy the given figure, and locate each of the following points.

a. Point *D*, such that $\overrightarrow{CD} = \overrightarrow{AB}$

b. Point *E*, such that $\overrightarrow{CE} = 2\overrightarrow{AB}$

c. Point *F*, such that $\overrightarrow{CF} = -\overrightarrow{AB}$

5. Given points $A(3, 1)$, $B(6, 5)$, and $C(4, -1)$, give the coordinates of the points D, E, and F described in Exercise 4.

6. $ABCD$ is a rhombus. Indicate which of the following are true, and which are false.

   a. $\overrightarrow{AB} = \overrightarrow{BC}$ **c.** $\overrightarrow{AB} + \overrightarrow{AD} = \overrightarrow{AC}$

   b. $\overrightarrow{AD} = \overrightarrow{BC}$

7. Using the given figure, complete the following equalities.

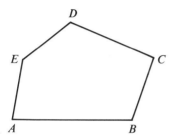

   a. $\overrightarrow{AB} + \overrightarrow{BC} = \_\_?\_\_$

   b. $\overrightarrow{ED} + \overrightarrow{DC} = \_\_?\_\_$

   c. $\overrightarrow{EB} + \overrightarrow{BC} = \_\_?\_\_$

   d. $(\overrightarrow{CA} + \overrightarrow{AE}) + \overrightarrow{ED} = \_\_?\_\_$

   e. $\overrightarrow{CA} + (\overrightarrow{AE} + \overrightarrow{ED}) = \_\_?\_\_$

8. The triangle method of adding arrows placed tip to end applies even when the arrows do not form a triangle. Using this information with points A, B, and C, shown in the figure below, complete the following equalities.

   a. $\overrightarrow{AB} + \overrightarrow{BC} = \_\_?\_\_$

   b. $\overrightarrow{CA} + \overrightarrow{AB} = \_\_?\_\_$

   c. $\overrightarrow{BA} + \overrightarrow{AC} = \_\_?\_\_$

9. Using the cube shown, name *all* arrows equivalent to each of the following.

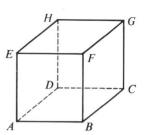

   a. $\overrightarrow{AB}$ **d.** $\overrightarrow{EF} + \overrightarrow{EH}$

   b. $\overrightarrow{GF}$ **e.** $\overrightarrow{EF} + \overrightarrow{ED}$

   c. $\overrightarrow{BG}$ **f.** $\overrightarrow{AC} + \overrightarrow{AE}$

10. Using the figure from Exercise 9 name an arrow equivalent to each of the following.

   a. $\overrightarrow{FG} + \overrightarrow{GH} + \overrightarrow{HD}$

   b. $\overrightarrow{AB} + \overrightarrow{AD} + \overrightarrow{AE}$

   c. $\overrightarrow{EH} + \overrightarrow{EF} + \overrightarrow{EA}$

   d. $\overrightarrow{AB} + \overrightarrow{BF} + \overrightarrow{FE} + \overrightarrow{EA}$

   e. Suggest a name for the arrow in Part d and for the translation it represents.

11. a. Sketch an arrow which represents the motion of an airplane flying east at 400 miles per hour.

   b. On the same sketch, draw an arrow representing a wind blowing from north to south at 100 miles per hour.

   c. Draw the arrow which represents the plane's movement when it encounters the wind. About how fast is the plane traveling? What is the approximate direction in which the plane is moving?

12. a. If translation T maps the origin to (2, 3), and translation S maps the origin to (5, 1), show that translation ST maps the origin to (7, 4). (Prove *ABDC* is a parallelogram.)

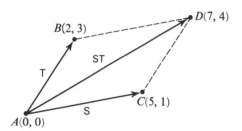

b. In Part a, T may be identified either by $\overrightarrow{AB}$ or by the ordered pair (2, 3). Likewise, S may be identified by $\overrightarrow{AC}$ or (5, 1). When ordered pairs are used, $\overrightarrow{AB} + \overrightarrow{AC} = \overrightarrow{AD}$ may be rewritten as (2, 3) + (5, 1) = (7, 4). Using this as an example, complete the following statement.

$$(a, b) + (c, d) = (\_\_?\_\_, \_\_?\_\_)$$

13. As was discussed in the previous exercise, the translation T which maps the origin to the point (6, 4) may be represented by the ordered pair of numbers (6, 4). Tell what ordered pair of numbers represents each of the following translations.

a. T^2 **b.** T^3 **c.** T^{-1} **d.** T^{-2}

(Note. T^{-2} is the inverse of T^2.)

14. The instructions for reaching a hidden treasure are as follows: Start at Dead Man's Cave. Walk 20 paces south, then 50 paces east, followed by 35 paces southwest.

a. Make a sketch using arrows to represent the three paths given in the instructions.

b. Draw the arrow representing the single path which will lead from the cave directly to the treasure. Determine the approximate length and direction of this path.

In Exercises 15–18, arrows will be identified by a single letter, such as $\vec{a}$, $\vec{b}$, and $\vec{x}$.

15. Using the diagram of $\vec{a}$, sketch arrows equivalent to $-\vec{a}$, $2\vec{a}$, $-2\vec{a}$, $3\vec{a}$ and $-3\vec{a}$.

$$\xrightarrow{\hspace{3cm}}$$
$$\vec{a}$$

16. Subtraction of arrows is defined in the following way.

$$\vec{a} - \vec{b} = \vec{a} + (-\vec{b})$$

Study the given figure, then draw arrow diagrams to represent $\vec{a} - 2\vec{b}$ and $\vec{a} - 3\vec{b}$.

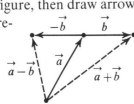

17. *ABCDEF* is the regular hexagon shown. Express each of the following in terms of $\vec{x}$, $\vec{y}$, and $\vec{z}$.

a. $\overrightarrow{AC}$ **c.** $\overrightarrow{AE}$ **e.** $\overrightarrow{DA}$

b. $\overrightarrow{CE}$ **d.** $\overrightarrow{BD}$

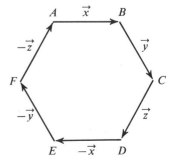

18. a. The figure shows the endpoints of several arrows from the origin. Copy the figure, and indicate the endpoints of $2\vec{a} - \vec{b}$, $2\vec{a} + 4\vec{b}$, and $-\vec{a} - \vec{b}$.

b. If $\overrightarrow{OP} = 2\vec{a} + 3\vec{b}$, P could be labeled (2, 3). The numbers 2 and 3 are called the coordinates of P relative to $\vec{a}$ and $\vec{b}$. Name the coordinates of Q, R, and S relative to $\vec{a}$ and $\vec{b}$.

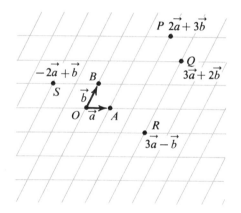

19. An airplane traveling north with a velocity of 300 miles per hour encounters a head wind of 60 miles per hour, blowing from the northeast. Draw an arrow diagram to represent this situation. Find the approximate speed of the plane, and the approximate direction in which the plane is moving.

20. A boy is headed south in a boat. He rows at 4 miles per hour, but the path of his boat is influenced by a current flowing from east to west at 3 miles per hour. Draw an arrow diagram of the situation. Determine the speed and the approximate direction of the boat.

21. Write a proof of Theorem 15 using the given figure. Let T be the translation which maps A to B and let S be the translation that maps B to C. Then, ST maps A to C. To prove that ST is a translation, show that ST maps any point, P, the same distance and direction as it maps A. That is, show that $\overrightarrow{PR} = \overrightarrow{AC}$.

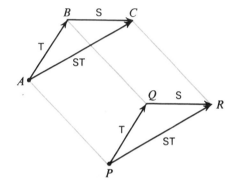

22. Write a proof of Theorem 16 using the given figure. Let translation T map A to B, and let translation S map B to C. Then, translation ST maps A to C. Show that TS also maps A to C. (Hint. Locate D so that $ABCD$ is a parallelogram.)

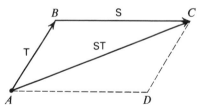

23. Let A, B, and P be three noncollinear points, and let $P' = H_B H_A(P)$. Prove that $\overrightarrow{PP'} = 2\overrightarrow{AB}$.

24. Describe the coordinate system that would result in Exercise 18 if $\vec{a} \perp \vec{b}$ and both were one unit long.

2.3   The Algebra of Half-Turns

We have previously stated that in many ways the composition of transformations resembles multiplication of real numbers. One fundamental property of multiplication of real numbers is:

If $a = b$ and $c = d$, then $ac = bd$,
where a, b, c, and d are real numbers.

The next theorem states that an analogous result holds for the composition of transformations.

THEOREM 17. If $S = T$ and $F = G$, then $SF = TG$, where S, T, F, and G are transformations.

Proof. To show that $SF = TG$, we must show that $SF(P) = TG(P)$ for every point P. Since $F = G$, $F(P) = G(P)$. Also, $S = T$, so $S(F(P)) = T(F(P))$. That is, $S(F(P)) = T(G(P))$ or $SF(P) = TG(P)$.

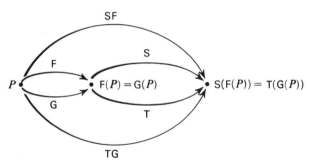

Figure 2.17

In Section 2.2, we proved that the product of translations is always a translation. In this section, we shall investigate products of half-turns. First, we shall state a useful theorem about half-turns. The proof follows immediately from Theorem 6.

THEOREM 18. A half-turn, H_p, is equivalent to a product of reflections in a pair of perpendicular lines which intersect at P.

For Theorems 19 and 20, both a geometric and an algebraic proof are given in an attempt to more closely relate the subjects of algebra and geometry.

THEOREM 19. The product of half-turns, $H_B H_A$, is equivalent to the translation represented by $2\overrightarrow{AB}$.

Geometric Proof. P is any point, with $P^* = H_A(P)$ and $P' = H_B H_A(P)$. Since A and B are the midpoints of $\overline{PP^*}$ and $\overline{P^*P'}$, respectively, $\overline{PP'}$ is parallel to $\overline{AB}$ and is twice as long as $\overline{AB}$. That is, $\overrightarrow{PP'} = 2\overrightarrow{AB}$. Also, Figure 2.18a illustrates that for any other point Q, $\overrightarrow{QQ'} = 2\overrightarrow{AB}$. Therefore, $\overrightarrow{PP'} = \overrightarrow{QQ'}$, and any two points are translated the same distance and direction by the product of the half-turns.

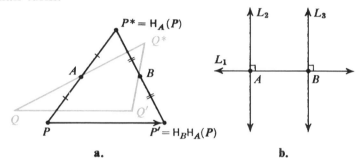

Figure 2.18

Algebraic Proof. By Theorem 18, each half-turn can be expressed as a product of reflections in perpendicular lines. Let $L_1 = \overleftrightarrow{AB}$. L_2 is the line perpendicular to L_1 at A, and L_3 is the line perpendicular to L_1 at B, as shown in Figure 2.18b. Then $H_A = R_1 R_2$ and $H_B = R_3 R_1$.

$$\begin{aligned}
H_B H_A &= (R_3 R_1)(R_1 R_2) \\
&= R_3 (R_1 R_1) R_2 \\
&= R_3 I R_2 \\
&= R_3 R_2
\end{aligned}$$

Since $L_2 \parallel L_3$, then by Theorem 8, $R_3 R_2$ equals a translation in the direction from L_2 to L_3 through twice the distance AB. This translation may be represented by $2\overrightarrow{AB}$. Therefore, $H_B H_A$ is equivalent to the translation represented by $2\overrightarrow{AB}$.

The proofs of the following corollaries are left as exercises.

COROLLARY 1 to THEOREM 19. If M is the midpoint of $\overline{AB}$, then the translation represented by $\overrightarrow{AB}$ equals $H_M H_A$ and $H_B H_M$.

COROLLARY 2 to THEOREM 19. $H_B H_A = H_C H_D$ if and only if $ABCD$ is a parallelogram.

THEOREM 20. If A, B, and C are noncollinear points, then $H_C H_B H_A = H_D$, where D is the point for which $ABCD$ is a parallelogram.

Geometric Proof. Let D be the point for which $ABCD$ is a parallelogram. If P is an arbitrary point, we must show that $H_C H_B H_A(P) = H_D(P)$. Let $Q = H_A(P)$, $M = H_B H_A(P)$ and $N = H_C H_B H_A(P)$, as shown in Figure 2.19.

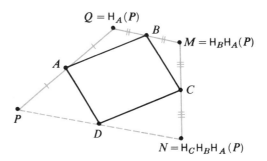

Figure 2.19

1. $\overline{DC} \parallel \overline{AB}$ because $ABCD$ is a parallelogram.
2. $\overline{AB} \parallel \overline{PM}$ because A and B are midpoints of $\overline{PQ}$ and $\overline{QM}$.
3. $\overline{DC} \parallel \overline{PM}$ because parallelism is transitive.
4. D is the midpoint of $\overline{PN}$, because $\overline{DC} \parallel \overline{PM}$ and C is the midpoint of $\overline{MN}$.
5. Therefore, $H_D(P) = N$.
6. But, $N = H_C H_B H_A(P)$, so $H_D(P) = H_C H_B H_A(P)$.

Since the result holds for any point P, $H_D = H_C H_B H_A$.

Algebraic Proof. If $ABCD$ is a parallelogram, then $H_B H_A = H_C H_D$ by Corollary 2 to Theorem 19.

$$\begin{aligned} H_C(H_B H_A) &= H_C(H_C H_D) \\ &= (H_C H_C) H_D \\ &= I H_D \\ &= H_D \end{aligned}$$

COROLLARY to THEOREM 20. The product of a half-turn and a translation is a half-turn.

Proof. By Theorem 19, a translation may be expressed as the product of two half-turns. Therefore, the product of a half-turn and a translation can be expressed as the product of three half-turns. By Theorem 20, this is equivalent to a single half-turn.

Theorem 19 discussed the product of two half-turns, and Theorem 20 discussed the product of three half-turns. Let us now consider products of four or more half-turns.

1. $H_D H_C H_B H_A = H_D(H_C H_B H_A)$
   $\qquad\qquad = H_D H_X \qquad$ (Theorem 20)
   $\qquad\qquad = \text{translation } T \quad$ (Theorem 19)

2. $H_E H_D H_C H_B H_A = H_E(H_D H_C H_B H_A)$
   $\qquad\qquad\qquad = H_E T \quad$ (By result 1)
   $\qquad\qquad\qquad = H_Y \quad$ (Corollary to Theorem 20)

3. $H_F H_E H_D H_C H_B H_A = H_F(H_E H_D H_C H_B H_A)$
   $\qquad\qquad\qquad\qquad = H_F H_Y \qquad$ (By result 2)
   $\qquad\qquad\qquad\qquad = \text{translation } S \quad$ (Theorem 19)

This pattern is summarized in the following theorem.

THEOREM 21. The product of an even number of half-turns is a translation. The product of an odd number of half-turns is a half-turn.

The following examples illustrate some interesting results of the theorems stated in this section.

Example 1. Show that when a point is reflected successively in the vertices of a parallelogram, the final image of the point is the same as the original point.

Solution. Let *ABCD* be the parallelogram shown in Figure 2.20. By Theorem 20, $H_C H_B H_A = H_D$. Theorem 17 allows us to "multiply" both sides of this equation by H_D, giving

$$H_D H_C H_B H_A = H_D H_D = I.$$

Therefore, as illustrated,

$$H_D H_C H_B H_A(P) = I(P) = P.$$

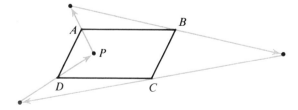

Figure 2.20

Example 2. Show that when a point is twice reflected successively in the vertices of a triangle, the final image of the point is the same as the original point.

Solution. Let ABC be the triangle shown in Figure 2.21. By Theorem 20, $H_C H_B H_A = H_D$. Theorem 17 allows us to "multiply" each side of this equation by itself. Therefore,

$$(H_C H_B H_A)(H_C H_B H_A) = H_D H_D.$$

That is, $\qquad H_C H_B H_A H_C H_B H_A = I.$

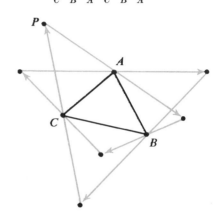

Figure 2.21

EXERCISES

1. Draw three noncollinear points, A, B, and P. Find $H_B H_A(P)$ and $H_A H_B(P)$.

2. Draw three noncollinear points A, B, and C. Locate point D so that $H_C H_B H_A = H_D$.

3. **a.** Draw a parallelogram, $ABCD$, and choose any arbitrary point, P. Reflect P in A, B, C, and D, to illustrate that $H_D H_C H_B H_A(P) = P$.

   b. How many quadrilaterals have A, B, C, and D as the midpoints of their sides?

4. **a.** Under what conditions does $H_C H_D = H_B H_A$?

   b. "Multiply" the equality in Part a by $H_D H_C$. Explain why the result simplifies to $I = H_D H_C H_B H_A$.

5. Draw a pentagon, $ABCDE$. Pick two arbitrary points, P and Q, and find P' and Q', their images by $H_E H_D H_C H_B H_A$. If you work accurately, the midpoint of $\overline{PP'}$ should be the same as the midpoint of $\overline{QQ'}$. Why?

6. a. Explain why $(H_E H_D H_C H_B H_A)^2 = I$.

   b. Illustrate Part a by taking an arbitrary point, P, and twice reflecting it successively in A, B, C, D, and E.

7. A, B, C, D, and E are the midpoints of the sides of pentagon $PQRST$, as shown. How could you construct the pentagon given only the points A, B, C, D, and E. (Hint. What is the image of P by $H_E H_D H_C H_B H_A$? In order to find P, pick any point X and locate $X' = H_E H_D H_C H_B H_A(X)$. P is the midpoint of $\overline{XX'}$. Why?)

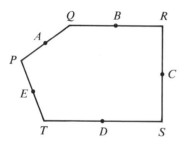

8. S and T are translations such that $S(A) = B$, $T(B) = C$. P and Q are the midpoints of $\overline{AB}$ and $\overline{BC}$, as shown. By Corollary 1 to Theorem 19, $S = H_B H_P$ and $T = H_Q H_B$. Prove that $TS = H_Q H_P$. (This is an alternate proof of Theorem 15.)

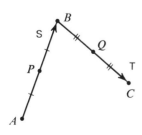

9. $L_1 \perp L_2$ at A and $L_1 \perp L_3$ at B. Prove $H_A R_3 = R_2 H_B$.

10. Prove Theorem 21 by considering two cases. If the number of half-turns is even, use Theorems 19 and 15. What will occur if the number of half-turns is odd?

11. Prove Corollary 1 to Theorem 19.

12. Prove Corollary 2 to Theorem 19.

13. Point A is on line L_1, as shown.

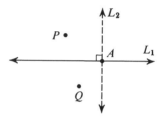

   a. Copy the given figure and find the images of P and Q by $H_A R_1$.

   b. What simple isometry equals $H_A R_1$? Prove this by letting $H_A = R_2 R_1$, where L_2 is the line perpendicular to L_1 at A.

14. A is a point not in line L_1. By Theorem 18, there are many pairs of lines, L_2 and L_3, such that $H_A = R_3 R_2$. Find one such pair for which you can prove that $H_A R_1$ is a glide reflection.

15. In a coordinate plane, $\triangle ABC$ has vertices $A(0, a)$, $B(0, 0)$, and $C(c, 0)$. Find the effect of $H_A H_B H_C$ on point $P(x, y)$. This product has the same effect as a half-turn about which point? Relate this result to Theorem 20.

16. A certain isometry has at least two fixed points. Show that it is a line reflection or the identity. If it has three noncollinear fixed points, show that it must be the identity.

2.4   The Algebra of Rotations

In Section 2.3 we considered the algebra of a special type of rotation, the half-turn. Recall that a half-turn equals a rotation of $180°$. For the larger class of all rotations, there are three important results we shall consider. The first of these asserts that a product of rotations like $C_{40}C_{30}$ is equivalent to C_{70}.

THEOREM 22. The product C_yC_x, of two rotations with the same center, is the rotation C_{x+y}.

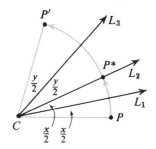

Figure 2.22

Proof. Let L_1, L_2, and L_3 be the lines shown in Figure 2.22. The angle of rotation from L_1 to L_2 is $\dfrac{x°}{2}$ and the angle of rotation from L_2 to L_3 is $\dfrac{y°}{2}$. By Theorem 6, $R_2R_1 = C_x$ and $R_3R_2 = C_y$.

$$C_yC_x = (R_3R_2)(R_2R_1)$$
$$= R_3(R_2R_2)R_1$$
$$= R_3R_1$$

Since the angle of rotation from L_1 to L_3 is $\dfrac{x}{2} + \dfrac{y}{2}$ degrees, by Theorem 6, $R_3R_1 = C_{x+y}$.

We have placed no restrictions on x and y in Theorem 22. Therefore, the theorem holds if x or y is negative or if $x + y \geq 360$. Figure 2.23 illustrates two such examples.

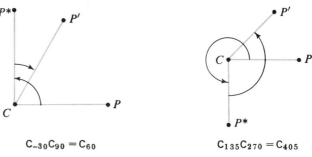

Figure 2.23

The next theorem considers the product of rotations with different centers.

THEOREM 23. The product $B_y A_x$, of two rotations with different centers, is a rotation or a translation.

$B_y A_x$ is a translation if $x + y$ is a multiple of 360.

$B_y A_x$ is a rotation if $x + y$ is not a multiple of 360.

Proof. L_1 is the line containing A and B. Let L_2 and L_3 be lines such that the angle of rotation from L_2 to L_1 is $\frac{x}{2}°$ and the angle of rotation from L_1 to L_3 is $\frac{y}{2}°$. By Theorem 6, $A_x = R_1 R_2$, and $B_y = R_3 R_1$.

$$B_y A_x = (R_3 R_1)(R_1 R_2)$$
$$= R_3(R_1 R_1)R_2$$
$$= R_3 R_2$$

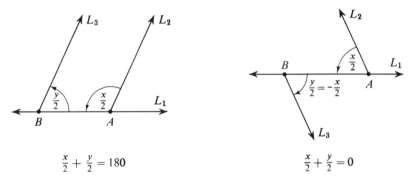

$$\tfrac{x}{2} + \tfrac{y}{2} = 180 \qquad\qquad \tfrac{x}{2} + \tfrac{y}{2} = 0$$

Figure 2.24

If L_2 and L_3 are parallel, by Theorem 8, $R_3 R_2$ is a translation. L_2 and L_3 will be parallel if $\frac{x}{2} + \frac{y}{2}$ equals 180 or a multiple of 180, including zero. (See Figure 2.24.) Therefore, when $x + y$ equals a multiple of 360, $B_y A_x$ is a translation.

If L_2 and L_3 intersect at some point C, by Theorem 6, $R_3 R_2$ is a rotation about C. (See Figure 2.25.) L_2 and L_3 will intersect if $\frac{x}{2} + \frac{y}{2}$ is not a multiple of 180. Therefore, when $x + y$ does not equal a multiple of 360, $B_y A_x$ is a rotation.

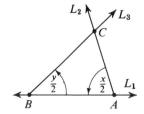

Figure 2.25

The last theorem of this section considers the product of a rotation and a translation. The proof is left as an exercise.

THEOREM 24. The product of a rotation and a translation is a rotation.

EXERCISES

1. Simplify the following products of rotations.

   a. $A_{120}A_{30}$ **c.** $(C_{90})^3$

   b. $A_{52}A_{-60}$ **d.** $(C_{90})^4$

2. Which of the following products are translations?

   a. $A_{120}B_{60}$ **d.** $A_{120}B_{120}C_{120}$

   b. $A_{240}B_{120}$ **e.** $A_{100}B_{120}C_{140}$

   c. $H_C H_D$

3. Write a proof for Theorem 24 for a translation T and a rotation A_x. Let $B = T(A)$. L_1 is the line perpendicular to $\overleftrightarrow{AB}$ at A, L_2 is the perpendicular bisector of $\overline{AB}$, and L_3 is the line such that the angle of rotation from L_3 to L_1 is $\dfrac{x}{2}^\circ$. Prove that TA_x is a rotation by writing T and A_x as products of reflections. Determine the magnitude and the center of the rotation TA_x.

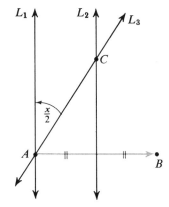

2.5 Groups

   Definition. A set S is **closed** under the operation * if and only if $a * b$ is an element of S, whenever a and b are elements of S.

Example 1.   **a.** The set of integers is closed under addition because the sum of any two integers is an integer.

        b. The set of integers is closed under subtraction because the difference of any two integers is an integer.

c. The set of positive integers is not closed under subtraction since the difference of two positive integers is not always a positive integer.

d. The set of odd integers is closed under multiplication but is not closed under addition.

e. The set of translations is closed under composition because the composite of two translations is always a translation. (Theorem 15)

f. The set of line reflections is not closed under composition because the composite of two reflections is not always a reflection.

We shall now discuss a pattern that appears frequently in mathematics.

Definition. A **group** consists of a set S and an operation $*$ which have the following properties.

1. *Closure:* If a and b are in S, then $a * b$ is in S.
2. *Associativity:* If a, b, and c are in S, then $(a * b) * c = a * (b * c)$.
3. *Existence of an Identity:* There is an identity element, I, in S such that $a * I = a = I * a$ for every a in S.
4. *Existence of Inverses:* If a is in S, then there is an inverse element in S, called a^{-1}, such that $a * a^{-1} = I = a^{-1} * a$.

Example 2. Show that R, the set of real numbers, with the operation of addition is a group.

Solution.
1. R is closed under addition, since the sum of two real numbers is always a real number.

2. From algebra, we know that the associative property holds for addition of real numbers.

3. There is a number, 0, in R such that $a + 0 = a = 0 + a$ for all a in R.

4. If a is in R, then there is an inverse, $-a$, in R such that $a + (-a) = 0 = (-a) + a$.

Example 3. Show that the set of translations with the operation of composition is a group.

Solution. 1. By Theorem 15, the composite of translations is a translation. Therefore the set of translations is closed under composition.

2. By Theorem 13, composition of transformations is associative. Therefore composition of translations is associative.

3. There is a translation, I, the zero translation, such that, for any translation, T, TI = T = IT.

4. If T is a translation, there is an inverse translation, T^{-1}, such that $TT^{-1} = I = T^{-1}T$. If $\overrightarrow{AB}$ represents T, then T^{-1} can be represented by either $-\overrightarrow{AB}$ or $\overrightarrow{BA}$.

Definition. A group is a **commutative group** if and only if it satisfies a fifth condition.

5. If a and b are in S, then $a * b = b * a$.

The group given in Example 2 is a commutative group because addition of real numbers is commutative. By Theorem 16, composition of translations is commutative. Therefore, the group given in Example 3 is also commutative.

Example 4. Is R', the set of nonzero real numbers, with the operation multiplication, a group? Is it a commutative group?

Solution. 1. R' is closed under multiplication.

2. Multiplication is associative in R'.

3. There is a number, 1, in R' such that $a \cdot 1 = a = 1 \cdot a$ for all a in R.

4. If a is in R', then there is an inverse, $\dfrac{1}{a}$, in R' such that $a\left(\dfrac{1}{a}\right) = 1 = \left(\dfrac{1}{a}\right)a.$

5. Multiplication is commutative in R'.

Therefore, R' with the operation of multiplication is a commutative group. (If we had considered the set of all real numbers instead, we would not have had a group, because zero has no multiplicative inverse.)

Example 5. Is the set of all transformations with the operation of composition a group? Is it a commutative group?

Solution. 1. By definition, the composite of two transformations is a transformation.

2. By Theorem 13, composition of transformations is associative.

3. There is a transformation, I, the identity mapping such that for any transformation, M, MI = M = IM.

4. If M is a transformation, then M is one-to-one and there is an inverse transformation, M^{-1}, such that $MM^{-1} = I = M^{-1}M$.

5. Composition of transformations is not commutative.

Therefore, the set of transformations with the operation of composition is a group, but is not a commutative group.

Three of the four examples of groups have been commutative groups. However, all four of the examples have been infinite groups, because the set of elements in each case was infinite. We now shall give two examples of finite groups. Since the group operations may not be familiar to you, we shall show in table form how the operation combines elements of the set.

Example 6. **a.** $S = \{0, 1, 2, 3, 4, 5\}$ operation denoted by *

*	0	1	2	3	4	5
0	0	1	2	3	4	5
1	1	2	3	4	5	0
2	2	3	4	5	0	1
3	3	4	5	0	1	2
4	4	5	0	1	2	3
5	5	0	1	2	3	4

$$4 * 2 = 0$$

b. $S = \{a, b, c, d, e, f\}$ operation denoted by ∘

∘	a	b	c	d	e	f
a	b	c	a	f	d	e
b	c	a	b	e	f	d
c	a	b	c	d	e	f
d	e	f	d	c	a	b
e	f	d	e	b	c	a
f	d	e	f	a	b	c

$$b \circ d = e$$

Let us check that each satisfies the four requirements of a group.

Closure: Since the elements in the table are only the elements of S, in each case, there is closure. If a 7 appeared in Table a or a k appeared in Table b, there would not be closure.

Associativity: It is a tedious job to check whether $(x * y) * z = x * (y * z)$ for all x, y, and z in S. For the present, we shall only illustrate that this holds for a special case, however it does hold in every case.

Table a: $(3 * 4) * 2 = 1 * 2 = 3$
$3 * (4 * 2) = 3 * 0 = 3$

Table b: $(b \circ d) \circ e = e \circ e = c$
$b \circ (d \circ e) = b \circ a = c$

Identity: The element 0 is the identity for Table a, and the element c is the identity for Table b.

Inverses: In Table a, 2 and 4 are inverses, since $2 * 4 = 0$, and 0 is the identity. Similarly, 1 and 5 are inverses because $1 * 5 = 0$. The elements 3 and 0 are their own inverses because $3 * 3 = 0$ and $0 * 0 = 0$. In Table b, a and b are inverses since $a \circ b = c$, and c is the identity. Each of the other four elements is its own inverse since the result of combining it with itself is the identity, c.

Thus, both examples are groups. Morever, Example 6a is a commutative group, and Example 6b is not a commutative group.

EXERCISES

1. Which of the following sets is closed under the given operation?
   a. Even integers; multiplication
   b. Integers; division
   c. Negative integers; addition
   d. Negative integers; multiplication
   e. All line reflections; composition
   f. All point reflections (half-turns); composition
   g. All half-turns and translations; composition
   h. All rotations; composition

2. Which of the following sets of transformations are closed under composition?
   a. All translations north
   b. All translations north or south
   c. All translations north, south, east, or west
   d. All translations
   e. All glide reflections
   f. All rotations with center C

3. Show that the group in Example 6b is not commutative.

4. Determine if each of the following sets is closed under addition, subtraction, multiplication, and division.

   a. Negative integers

   b. Perfect squares, $\{1, 4, 9, \ldots\}$

   c. Rational numbers

   d. Nonzero rational numbers

   e. Fractions with an even numerator and an odd denominator

5. In Example 6a, study the row of the table opposite the identity element, 0, and in Example 6b, study the row opposite the identity element, c. What do you notice? Why must this be true? Do you see any special column in either table?

6. The set $S = \{a, b, c, d\}$ is a group with two different operations, $*$ and $\sim$. Tables for these operations are shown below.

$*$	a	b	c	d
a	b	c	d	a
b	c	d	a	b
c	d	a	b	c
d	a	b	c	d

$\sim$	a	b	c	d
a	a	b	c	d
b	b	a	d	c
c	c	d	a	b
d	d	c	b	a

   a. What is the identity element for the operation $*$?

   b. Name the inverse of each element in the group with operation $*$.

   c. What is the identity element for the operation $\sim$?

   d. Name the inverse of each element in the group with operation $\sim$.

7. Show that the set of integers is a group with addition but not with multiplication.

8. Is the set of all even integers a group with addition? What about the set of all odd integers? Explain your answer.

9. a. Let S be the set of all multiples of 3, $\{\ldots, -6, -3, 0, 3, 6, \ldots\}$. Show that S is a group with addition.

   b. Is the set of all multiples of 4 a group with addition?

10. A partial table for a commutative group with 3 elements is given. Copy and complete the table. What is the identity element? Name the inverses.

$*$	a	b	c
a	?	a	?
b	?	?	c
c	b	?	?

11. From algebra, we have the following cancellation laws.

   1. If $a + b = a + c$, then $b = c$.

   2. If $ab = ac$ and $a \neq 0$, then $b = c$.

Here is a proof of cancellation law 1.

$$a + b = a + c$$
$$(-a) + (a + b) = (-a) + (a + c)$$
$$(-a + a) + b = (-a + a) + c$$
$$0 + b = 0 + c$$
$$b = c$$

   a. Prove cancellation law 2.

   b. Prove that if $H_A H_B = H_A H_C$, then $H_B = H_C$.

   c. The elements a, b, and c belong to set S, which is a group with the operation $*$. Prove the following cancellation law for groups.
      If $a * b = a * c$, then $b = c$.

   d. Explain why no row of a group table can have an element repeated.

2.6   Transformation Groups

When the elements of a group are transformations and the group operation is composition, the group is a transformation group. Example 3 of Section 2.5, the set of all translations, is a transformation group. The set of all rotations with the same center is also a transformation group. Two more transformation groups are considered in Theorems 25 and 26.

THEOREM 25. The set of all isometries is a transformation group.

Proof. 1. The composite of any two isometries is an isometry. (You will be asked to prove this in the exercises.)

2. Since every isometry is a transformation, by Theorem 13, the composition of isometries is associative.

3. The identity mapping, I, preserves distance so it is an isometry. Therefore, the set of all isometries has an identity element.

4. Every isometry has an inverse mapping that also preserves distance, so the inverse belongs to the set of all isometries.

Therefore, the set of all isometries is a transformation group.

The set of all isometries consists of rotations, translations, reflections, and glide reflections. We shall now consider a subset of the set of isometries consisting of all rotations and translations.

THEOREM 26. The set of all rotations and translations is a transformation group.

Proof. 1. By Theorem 15, the composite of two translations is a translation. By Theorems 22 and 23, the composite of two rotations is a rotation or a translation. By Theorem 24, the composite of a translation and a rotation is a rotation. Therefore this set is closed under composition.

2. By Theorem 13, composition is associative.

3. The identity mapping, I, may be thought of as a zero rotation or a zero translation. Therefore, I is in the set.

4. The inverse of a rotation is a rotation and the inverse of a translation is a translation. Therefore, every inverse is in the set.

Therefore, the set of all rotations and translations is a transformation group.

Beside the group of isometries, there is one other important group of transformations we shall consider. This is the group of transformations which map a figure to a similar figure. One of the simplest kinds of similarity transformations is a dilation, as shown in Figure 2.26. The center of the dilation is point O, and every point, P, is projected along $\overleftrightarrow{OP}$ to an image point, P', so that $\overrightarrow{OP'} = k\overrightarrow{OP}$, where k is a nonzero constant. When $k = 1$, the dilation is just the identity and when $k = -1$, it is a half-turn. Sometimes k is called the scale factor of the dilation. All dilations and all similarity transformations have the following properties.

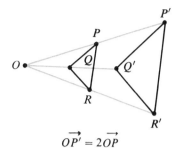

$$\overrightarrow{OP'} = 2\overrightarrow{OP}$$

1. They preserve ratios of distance: $\dfrac{PQ}{QR} = \dfrac{P'Q'}{Q'R'}$

2. They preserve angle measure: $m\angle PQR = m\angle P'Q'R'$

3. They preserve perpendicularity and parallelism.

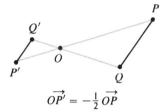

$$\overrightarrow{OP'} = -\tfrac{1}{2}\overrightarrow{OP}$$

Figure 2.26

The dilation with center at the origin of a coordinate plane can be defined by $\mathsf{D}:(x, y) \rightarrow (kx, ky)$. When $k > 0$, this dilation stretches the plane by a factor of k, in both the x and y directions. A more general transformation, $\mathsf{M}:(x, y) \rightarrow (ax, by)$, stretches the plane by different factors in the x and y directions. In Figure 2.27, $\triangle ABC$ is mapped to $\triangle A'B'C'$ by the transformation $\mathsf{M}:(x, y) \rightarrow \left(2x, \dfrac{1}{3}y\right)$. Such a transformation does not preserve angle measure, or ratios of distances. This type of mapping does, however, map lines to lines and midpoints of segments to midpoints of segments. Some transformations, however, do not even preserve these properties. Figure 2.28 illustrates the bending and twisting which occurs when a figure, F, is mapped to a figure, F', by a topological transformation.

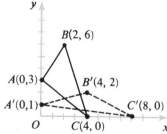

Figure 2.27

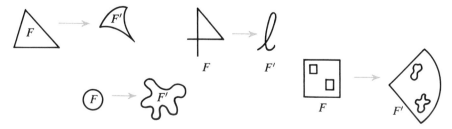

Figure 2.28

From the mappings in Figure 2.28, you may have surmised that topological transformations preserve few geometric properties. However, there are some properties which are preserved. For example, a figure with a number of holes in it is mapped to a figure with the same number of holes. Two figures with the same number of holes are said to be topologically equivalent because one can be mapped to the other by a topological transformation. This is analogous to calling figures congruent if one can be mapped to the other by an isometry, or similar if one can be mapped to the other by a similarity transformation. Moreover, the counterpart of saying "congruent figures have equal corresponding distances" is saying "topologically equivalent figures have an equal number of holes."

Whether we study the group of isometries, similarity transformations, or topological transformations, our prime interest is with the geometric properties of figures which are preserved by the group of transformations. The relationship among the groups of transformations we have mentioned is summarized in this chart.

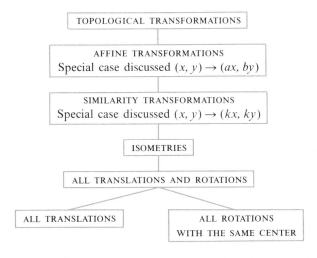

As you read from the top to the bottom of the chart, notice that the transformations preserve increasingly more properties.

If a group G consists of a set S and an operation *, then a *subgroup* of G consists of a subset of S which is a group under *. In the chart above, each group listed is a subgroup of the groups listed above it. That is, the set of affine transformations is a subgroup of the set of topological transformations, and the set of similarity transformations is a subgroup of both the set of affine transformations and the set of topological transformations.

EXERCISES

1. Verify that the set of all rotations with the same center is a transformation group.

2. For each of the following scale factors, sketch a triangle and its image under a dilation having its center at the centroid of the triangle.

    a. $\dfrac{1}{2}$ **b.** $-\dfrac{1}{2}$

3. D_3, D_2, and $D_{\frac{1}{2}}$ are dilations with the center O and scale factors of 3, 2, and $\dfrac{1}{2}$, respectively.

    a. Describe the mappings D_3D_2 and D_2D_3.
    b. Describe the mapping $D_2D_{\frac{1}{2}}$. What is the inverse of D_2?
    c. Prove that the set of all dilations with center O is a commutative group.

4. Demonstrate that the set of isometries is closed under composition.

5. For the dilation shown at the top of Figure 2.26, prove each of the following.

    a. $\overleftrightarrow{PQ} \parallel \overleftrightarrow{P'Q'}$
    b. $m\angle PQR = m\angle P'Q'R'$
    c. $\dfrac{PQ}{QR} = \dfrac{P'Q'}{Q'R'}$

6. Consider $\text{M}:(x, y) \rightarrow \left(2x, \dfrac{1}{3}y\right)$, the transformation shown in Figure 2.27.

    a. Prove that the midpoints of $\overline{AB}$, $\overline{BC}$, and $\overline{AC}$ are mapped to the midpoints of $\overline{A'B'}$, $\overline{B'C'}$, and $\overline{A'C'}$, respectively.
    b. If M maps lines to lines, explain why it must map the medians of $\triangle ABC$ to the medians of $\triangle A'B'C'$.
    c. What is the image of the centroid of $\triangle ABC$? Explain your answer. (Recall that the centroid of a triangle is the point of intersection of its medians.)

7. Which of the following figures are topologically equivalent?

    a.   **c.**   **e.**   **g.**   **i.**

    b.   **d.**   **f.**   **h.**   **j.**

2.7 Symmetry Groups

The group we shall discuss in the following example is called a symmetry group. Recall that a symmetry of a figure is an isometry which maps the figure onto itself.

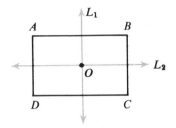

Figure 2.29

Example. The rectangle $ABCD$, shown in Figure 2.29, has four symmetries. Beside the identity map I there is a half-turn H_O and the reflections in L_1 and L_2.

First, note that $R_1R_1 = R_2R_2 = H_OH_O = I$. Also, by Theorem 18, $R_1R_2 = H_O = R_2R_1$. Figure 2.30 illustrates that $R_1 = R_2H_O$, and that $R_2 = H_OR_1$.

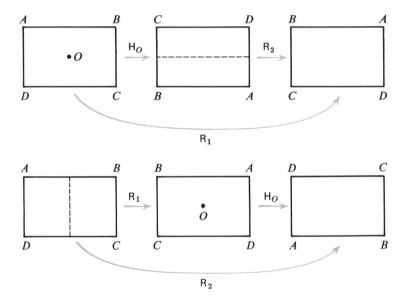

Figure 2.30

The following is the table for the symmetry group of the rectangle.

	I	H_O	R_1	R_2
I	I	H_O	R_1	R_2
H_O	H_O	I	R_2	R_1
R_1	R_1	R_2	I	H_O
R_2	R_2	R_1	H_O	I

In the exercises you will be asked to show that with composition, the symmetries of a rhombus, square, and equilateral triangle are also groups. Sometimes the symmetry group of a figure is small, as is the case with an isosceles triangle. Sometimes a group of symmetries is trivial, because only the identity symmetry maps the figure onto itself. However, the symmetries of any figure form a group.

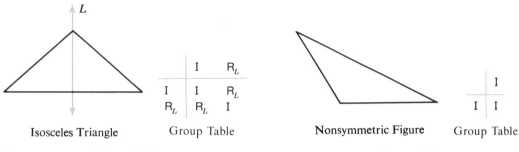

Isosceles Triangle	Group Table	Nonsymmetric Figure	Group Table
Figure 2.31		**Figure 2.32**	

THEOREM 27. The set of symmetries of any figure forms a group.

Proof. By definition, every symmetry is an isometry which maps a figure, F, onto itself.

1. Let S and T be symmetries of a figure F. Then, $S(F) = F$ and $T(F) = F$, therefore $ST(F) = S(T(F)) = S(F) = F$, and ST is a symmetry of F. Thus, the set of symmetries of a figure is closed under composition.

2. Since composition of transformations is associative by Theorem 13, composition of symmetries is associative.

3. The identity mapping I is a symmetry of F, since $I(F) = F$.

4. If T is a symmetry of F, it is an isometry which maps F onto itself. But, every isometry has an inverse which, in this case, also maps F onto itself. Consequently, the inverse of a symmetry of F is also a symmetry of F.

Thus, the set of symmetries of F is a group with composition.

Although we have illustrated the result of Theorem 27 for a rectangle and an isosceles triangle, there is nothing in the proof which requires the figure to be finite. This theorem applies equally to infinite figures. Figure 2.33 shows a series of patterns which we can imagine to be extended indefinitely to the left and right. One symmetry of the figure is the translation T which maps each pattern one "unit" to the right. T maps the extended figure onto itself. The translations T^2, T^3, T^4, ..., and T^{-1}, T^{-2}, T^{-3}, ... also map the figure onto itself, as does the identity, $T^0 = I$. Thus, the symmetry group of the figure is $\{T^n: n$ is an integer$\}$.

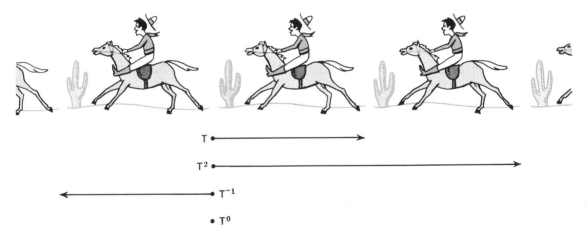

Figure 2.33

Figure 2.34 shows another figure which we can imagine to be extended indefinitely in all directions, to fill the plane. It is from *The Graphic Work of M. C. Escher*. The symmetries of this figure are all translations. Let T be the translation that maps each fish one "unit" right, and S be the translation that maps each fish one "unit" up. Then, TS maps each fish one unit right and one unit up, and T^2S^{-1} maps each fish two units right and one unit down. Thus, the symmetry group of this figure is $\{T^mS^n: m$ and n are integers$\}$. When m and n are both zero, T^mS^n is the identity.

Finally, the topic of symmetry is easily generalized to three dimensions. The Eiffel Tower, for example, has a total of eight symmetries. Four are reflections in a plane, three are rotations about the axis of the tower and the eighth is the identity. For this structure, as for any structure, the set of all symmetries is a transformation group.

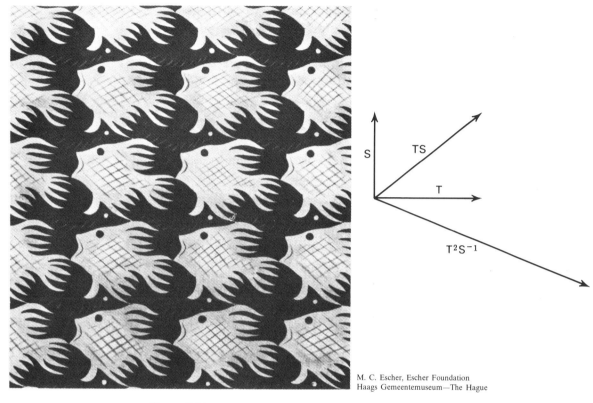

M. C. Escher, Escher Foundation
Haags Gemeentemuseum—The Hague

Figure 2.34

EXERCISES

1. Rhombus *ABCD* has four symmetries. Determine what they are and make a table similar to that in the example of this section. Is your table a group table? Is it commutative?

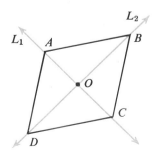

2. The two figures given have only rotational symmetry. The figure at the left is the triquetrum, an ancient magic symbol. Make group tables for the symmetries of both figures. Don't forget that the identity mapping is a symmetry of both figures.

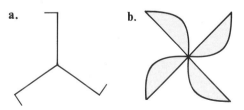

3. Equilateral triangle ABC has three line symmetries, R_1, R_2, and R_3, the rotational symmetries, O_{120} and O_{240}, and the identity I. Make a group table for this figure. Here are three examples to help you fill it out.

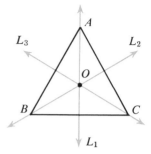

a. $R_1 R_2$ is the composite of reflections in intersecting lines, and is equal to a rotation of twice the angle from L_2 to L_1. Therefore, $R_1 R_2 = O_{-240} = O_{120}$.

b. Similarly, $R_2 R_3$ equals a rotation of twice the angle from L_3 to L_2, or $R_2 R_3 = O_{120}$.

c.

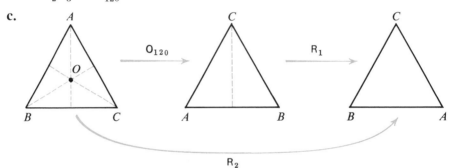

Thus, $R_1 O_{120} = R_2$.

4. A square has four line symmetries, three rotational symmetries, and the identity. Make the group table for this figure. Is it a commutative group?

5. We can imagine the figure shown on page 45 to be extended indefinitely in all directions to fill the plane. Its symmetries are glides and glide reflections. Describe these symmetries. (Ignore the color distinction between the birds.)

6. Describe the symmetries of the figures in Exercise 15 of Section 1.8.

Selected Answers

Section 1.1, Pages 3-4.

2. $\frac{1}{2}$; no

3. a. They are the same size. **b.** The distance on the globe is greater than the distance on the map. The ratio of distances is $\frac{\pi}{2}$.

4. a. N **b.** A circle **c.** No **d.** No

5. a. No; Yes **b.** No

6. a. A square **b.** A square **c.** A circle contained in the top of the box **d.** The North Pole is its own image. **e.** Neither **f.** Regions near the 6 points of tangency of the box and the sphere

Section 1.2, Pages 7-8.

1. a. Domain: L_1; Range: L_2 **b.** Yes **c.** Yes

2. b. The distance from P' to Q' is twice the distance from P to Q. **c.** Both the domain and the range are the entire plane. **d.** Yes **e.** Yes

3. a. $(-3, 0), (\frac{9}{2}, 3)$ **b.** $(\frac{2}{3}, 1), \left(\frac{a}{3}, \frac{b}{3}\right)$ **c.** The distance from P' to Q' is 3 times the distance from P to Q.

4. a. For any point P in $\overline{AB}$, let the image of P be the point of intersection of $\overline{CD}$ with the line through P which is parallel to $\overline{BC}$ and $\overline{AD}$. **b.** Answers will vary. One possibility is for every point P of L_1 let P' be the projection of P in L_2.

5. The image is $\triangle A'B'C'$ with vertices $A'(0, 0)$, $B'(1, 5)$, and $C'(3, 6)$. M preserves distance.

6. The image is $\triangle A'B'C'$ with vertices $A'(-1, 6)$, $B'(4, 2)$, and $C'(2, 0)$. M preserves neither distance nor shape.

7. The image is $\triangle A'B'C'$ with vertices $A'(-2, 6)$, $B'(8, 2)$, and $C'(4, 0)$. M does not preserve distance. M does preserve the shape of $\triangle ABC$.

8. The image is $\triangle A'B'C'$ with vertices $A'(1, -3)$, $B'(-4, -1)$, and $C'(-2, 0)$. M preserves distance. M preserves the shape of $\triangle ABC$.

9. b. They are the same segment. **c.** The image of the intersections is two points of L. The intersection of the images is a segment of L.

Section 1.3, Page 11.

1. a. $F(-1) = -1, F(0) = 1, F(1) = 3$, and $F(2) = 5$; Range: set of odd integers; Yes **b.** $F(-1) = -5$, $F(0) = 0, F(1) = 5$, and $F(2) = 10$; Range: set of all multiples of 5; Yes **c.** Range: I; Yes

2. a. $F(0) = 5, F(1) = 8, F(-1) = 8$, and $F(2) = 17$ **b.** The set of all real numbers greater than or equal to 5 **c.** No **d.** $\pm 10, \pm \sqrt{7}$

3. Answers will vary. One possibility is:
a. $X \rightarrow$ mother of X **b.** $X \rightarrow$ weight of X.

4. a. The slope of a vertical line is undefined. **b.** The set of all real numbers **c.** No

5. When x is an integer, $\frac{x}{2}$ will not always be an integer. However, if x is a real number, $\frac{x}{2}$ is always a real number.

6. a. 5 **b.** Many answers are possible, including $(0, 5)$, $(-5, 0), (4, 3)$ and $(-3, 4)$.

7. a. $(10, -3), (0, 2)$ **b.** $\left(\frac{7}{2}, 1\right), (0, -5), \left(\frac{a}{2}, b\right)$ **c.** The domain and the range are both the set of all points of the plane. **d.** Yes; Yes **e.** When P and Q have the same x-coordinate, distance is preserved. M does not preserve distance.

8. a. $(4, -8), (-3, 4), (1, 0)$ **b.** Domain: all points of the plane; Range: $\{(x, y) | (x < -1) \text{ or } (x \geq 1)\}$ **c.** Yes; No **d.** When P and Q have the same x-coordinate, distance is preserved. M does not preserve distance.

Section 1.4, Pages 15-16.

4. a. TOT, MUM, WOW **b.** Any word composed solely of the letters B, C, D, E, H, I, O, and X will have this property.

5. SOS remains the same when reflected in a point.

7. a. In each case, the line $y = x$ contains the midpoint of the segment. Also, the slope of the line is 1 and the slope of each segment is -1. **b.** Let (a, b) be any point, and let its image by a reflection in the line $y = x$ be the point (c, d). The line $y = x$ contains $\left(\frac{a + c}{2}, \frac{b + d}{2}\right)$, so $\frac{a + c}{2} = \frac{b + d}{2}$, or $a + c = b + d$. Also, the slope of the segment, $\frac{d - b}{c - a}$, must equal -1, so $d - b = a - c$. Solving these equations we find $a = d$ and $b = c$, so $(c, d) = (b, a)$. **c.** M:$(a, b) \rightarrow (-b, -a)$

8. a. $(5, -2); (-5, 2); (2, 5); (-5, -2)$ **b.** $(-2, 0); (2, 0); (0, -2); (2, 0)$ **c.** $(1, -\sqrt{6}); (-1, \sqrt{6}); (\sqrt{6}, 1); (-1, -\sqrt{6})$

9. a. $(-a, -b)$ **b.** R_x:$(a, b) \rightarrow (a, -b)$ and R_y:$(a, -b) \rightarrow (-a, -b)$

10. a. $(0, 2)$ **b.** $(4, 4)$ **c.** $(5, -1)$ **d.** $(-1, 7)$ **e.** $(0, 4)$ **f.** $(5, 7)$

11. a. $y = -x$ **b.** $y = -x + 10$

12. a. $(2, 3)$ **b.** $(0, 3)$ **c.** $\left(\frac{x_1 + x_2}{2}, \frac{y_1 + y_2}{2}\right)$

13. a. $(4, 4)$ **b.** $x + y = -4$

14. *Case a.* By the definition of line reflection, $\overline{QN} \cong \overline{Q'N}$, $\overline{PM} \cong \overline{P'M}$, $\overline{MN} \perp \overline{PP'}$, and $\overline{MN} \perp \overline{QQ'}$. $\triangle QNM \cong \triangle Q'NM$ by the SAS congruence, therefore $\overline{QM} \cong \overline{Q'M}$ and $\angle QMN \cong \angle Q'MN$. Complements of congruent angles are congruent, so $\angle PMQ \cong \angle P'MQ'$. By the SAS congruence, $\triangle PMQ \cong \triangle P'MQ'$. Therefore, $PQ = P'Q'$.
Case b. Since P and Q are contained in L, $P = P'$ and $Q = Q'$. Therefore $PQ = P'Q'$.
Case c. Q is contained in L, therefore $Q = Q'$. Since $\overline{PQ} \perp L$, Q is the midpoint of $\overline{PP'}$. Thus, $PQ = P'Q$, or $PQ = P'Q'$.
Case d. Q is contained in L, therefore $Q = Q'$. Since Q lies in the perpendicular bisector of $\overline{PP'}$, $PQ = P'Q$. Thus, $PQ = P'Q'$.
Case e. The proof of this case is similar to the proof of case a.

15. *Case a.* Since $PO = P'O$ and $QO = Q'O$, $\triangle QOP \cong \triangle Q'OP'$ by the SAS congruence, and $PQ = P'Q'$.
Case b. $PO = P'O$ and $QO = Q'O$, therefore $PO - QO = P'O - Q'O$. That is $PQ = P'Q'$.
Case c. Since $P = O$, $P = P'$. Also, $OQ = OQ'$, therefore $PQ = P'Q'$.

16. **a.** $P'Q' = \sqrt{[(x_2 + 3) - (x_1 + 3)]^2 + (y_2 - y_1)^2}$
$= \sqrt{(x_2 - x_1)^2 + (y_2 - y_1)^2} = PQ$

b. $P'Q' =$
$\sqrt{[(x_2 + 3) - (x_1 + 3)]^2 + [(y_2 - 2) - (y_1 - 2)]^2}$
$= \sqrt{(x_2 - x_1)^2 + (y_2 - y_1)^2} = PQ$

17. a. No **b.** Area of $\triangle QNR$ = area of $\triangle Q'NR'$, area of $\triangle QPN$ = area of $\triangle Q'P'N$, and area of $\triangle PMN$ = area of $\triangle P'MN$ because in each case the triangles have equal bases and the same altitude. Therefore, area of $PQRM$ = area of $P'Q'R'M$. That is, area of $\triangle PQR$ + area of $\triangle PMR$ = area of $\triangle P'Q'R$ + area of $\triangle PMR$. Since area of $\triangle PMR$ = area of $\triangle P'MR'$, then area of $\triangle PQR$ = area of $\triangle P'Q'R'$. **c.** Yes. Let S and T be midpoints of $\overline{PQ}$ and $\overline{P'Q'}$. Then $\overline{ST}$ is the median of trapezoid $PQQ'P'$. Use the properties of a median of a trapezoid to show that $\overline{ST}$ is parallel to L_2 and bisected by L_1. Thus, T is the image of S.

Section 1.5, Pages 20–22.
2. The distance along the river to the pumping station from the perpendicular through A should be 4 miles. The amount of pipe needed is $5\sqrt{5}$ miles.
4. Let the incoming ray form an angle of $x°$ with the mirror. The reflected ray also forms an angle of $x°$ with the mirror, but it forms an angle of $(90 - x)°$ with the second mirror. The emerging ray, therefore, forms an angle of $(90 - x)°$ with the second mirror. Since a pair of interior angles on the same side of a transversal are supplementary, the initial and final rays are parallel.

5. This is true for any point $P(0, y)$ on $\overline{OD}$.
6. Find the image of the line by a half-turn about P. Let either point of intersection of the triangle and the image of the line be A and call its preimage B.
7. Find the image of C_1 by a half-turn about P. The image of C_1 will intersect C_2 in P and a second point Q. $\overleftrightarrow{PQ}$ is the desired line.
8. $X(5, 0)$, $Y(4, 4)$, and $Z(3\frac{1}{3}, 4)$
9. a. Reflect B in L_1 to get $B'(16, 9)$ and then reflect B' in L_2 to get $B''(16, -9)$. The shortest path will intersect L_2 at $(4, 0)$ and then intersect L_1 at $(10\frac{2}{3}, 5)$. The length of this path will equal AB'' or 20 units.
b. Reflect B in L_2 to get $B'(16, -1)$ and then reflect B' in L_1 to get $(16, 11)$. The shortest path will intersect L_1 at $(4, 5)$ and then L_2 at $(14, 0)$. The length of this path will equal AB'' or $8\sqrt{5}$ units.
10. Find the image of C_2 by a reflection in L. The image of C_2 will intersect C_1 in two points. Call either of these points Q. Let the preimage of Q be S. $\overline{QS}$ is a diagonal of the desired square. Find P and R on L so that $PR = QR$, and $\overline{PR}$ and $\overline{QS}$ bisect each other. There are two possible squares.
11. Find C_1', the image of C_1 by a reflection in L. C_1' and C_2 have four common tangents. One of these tangents will intersect L in P. Each of the other three tangents will intersect L in a point which has the required property.

Section 1.6, Pages 24–25.

1. Let M be the midpoint of $\overline{AB}$, and let M' be its image. Since the mapping is an isometry, A', B', and M' are collinear, and $AM = A'M'$ and $BM = B'M'$. Since $AM = BM$, $A'M' = B'M'$, and M' is the midpoint of $\overline{A'B'}$.
2. Perpendicular lines form a 90° angle. Since an isometry preserves angle measure, the angle formed by the images of the lines is also a 90° angle. Therefore, the image lines are also perpendicular.
3. a. Yes **b.** Yes **c.** Yes **d.** The distance between any two points is half the distance between their images.
4. a. No **b.** M preserves collinearity, parallelism, and orientation. M does not preserve angle measure.
c. Consider $A(x_1, y_1)$ and $B(x_2, y_2)$ and their images by M, $A'(x_1, 2y_1)$ and $B'(x_2, 2y_2)$. The midpoint of $\overline{A'B'}$ is the point with coordinates $\left(\frac{x_1 + x_2}{2}, y_1 + y_2\right)$. The preimage of this point has coordinates $\left(\frac{x_1 + x_2}{2}, \frac{y_1 + y_2}{2}\right)$ and is therefore the midpoint of $\overline{AB}$.
5. a. F and F'' have the same orientation. **b.** In Britain, cars drive on the left-hand side of the road, and the steering wheel is on the right-hand side of the car.

6. Let L_1 and L_2 be parallel lines. Then, $L_1 \cap L_2 = \phi$. Therefore, $L_1' \cap L_2' = \phi$. That is, L_1' and L_2' are parallel.

Section 1.7, Pages 29–31.

2. a. D **b.** B **c.** C **d.** A

3. a. P **b.** R **c.** R **d.** Q

4. a. By the definition of rotation, $OA = OA'$. Therefore, O lies in the perpendicular bisector of $\overline{AA'}$.

5. a. B_{-60} **b.** A and Y **c.** Theorem 5 **d.** $60°$ (See Example 1.)

6. a. Let the three parallel lines be called L_1, L_2, and L_3, with L_2 between L_1 and L_3. Choose an arbitrary point A on L_1 and rotate L_2 about A through $60°$. Let C be the intersection of the image line with L_3. If B is the preimage of C, then $\triangle ABC$ is the desired triangle. There are two solutions since L_2 can be rotated $60°$ in either direction. **b.** Let the three circles be called C_1, C_2, and C_3, with C_2 between C_1 and C_3. Choose an arbitrary point A on C_1 and proceed as in part a.

7. a. C_{90} **b.** A and D **c.** Theorem 5 **d.** The angle between $\overleftrightarrow{AD}$ and $\overleftrightarrow{BE}$ is the same as the angle of rotation, $90°$. (See Example 1.)

8. a. C_{-240}, C_{480} **b.** C_{-430}, C_{290} **c.** C_{-149}, C_{571} **d.** C_{-640}, C_{-280}, C_{80} **e.** C_{-360}, C_{360} **f.** C_{-180}, C_{540} (Other answers are possible.)

9. Find L', the image of L by a rotation of $90°$ about A. L' intersects C in two points. Call either point Z, and let the preimage of Z be X. Then, find Y so that $YZ = YX = AX$.

11. Solution is similar to that of Exercise 6b. 0, 1, or 2 points of intersection are possible depending on the size of the circles and the location of A.

12. $\angle PCP'' = \angle PCA + \angle ACB + \angle BCP''$. Since reflections preserve angle measure, $\angle PCA = \angle ACP'$ and $\angle P'CB = \angle BCP''$. Therefore, $\angle PCP'' = \angle ACP' + \angle ACB + \angle P'CB$. Since $\angle ACB = \angle ACP' + \angle P'CB$, then $\angle PCP'' = 2\angle ACB$. But, $\angle ACB = x$, so $\angle PCP'' = 2x$.

In Figure 1.39b, $\angle PCP'' = 180 - (\angle DCP + \angle P''CB)$. $\angle DCP = \angle P'CD = 180 + (\angle P'CA + \angle ACB)$. $\angle P'CA = \angle ACP'' = x + y$, so $\angle DCP = 180 + (2x + y)$. Therefore, $\angle PCP'' = 180 - [180 + (2x + y) - y] = -2x$.

Section 1.8, Pages 34–36.

2. $(6, 2)$, $(8, -5)$, $(10, 4)$, and $(0, 0)$

3. a. $(-1, -4)$, $(8, -6)$, and $(x + 4, y - 3)$ **b.** $(-4, 3)$

4. a. $(4, 6)$ and $(0, 4)$ **b.** 5 units

5. The midpoints of $\overline{AA'}$, $\overline{BB'}$, and $\overline{CC'}$ are contained in the line of reflection, L. Let A^* be the glide image of A. Consider the triangle AA^*A'. L contains the midpoint of $\overline{A^*A'}$ and is parallel to $\overleftrightarrow{AA^*}$. Therefore, L contains the midpoint of $\overline{AA'}$.

7. Find C_1', the image of C_1 by the translation $\overrightarrow{AB}$. C_1' and C_2 intersect in two points. Call either point X, and call its preimage Y.

8. Find L_1', the image of L_1 by the translation $\overrightarrow{AB}$. L_1' and L_2 will intersect in a point. Call this point X, and call its preimage Y.

9. a. A translation or a rotation (half-turn) **b.** A reflection **c.** A glide reflection **d.** A translation or a rotation (half-turn) **e.** A translation or a rotation (half-turn)

10. a. A rotation **b.** A rotation (half-turn) **c.** A translation **d.** A glide reflection **e.** A rotation

11. a. L_1 is the perpendicular bisector of $\overline{PP'}$, so the distance from P' to L_1 is $\frac{1}{2}PP'$. Also, L_2 is the perpendicular bisector of $\overline{P'P''}$, so the distance from P' to L_2 is $\frac{1}{2}P'P''$. Therefore, $x = \frac{1}{2}PP' + \frac{1}{2}P'P''$, or $2x = PP' + P'P'' = PP''$. We know that P, P', and P'' are collinear because $\overleftrightarrow{PP'} \perp L_1$ and $L_1 \parallel L_2$, therefore $\overleftrightarrow{PP'} \perp L_2$. But, $\overleftrightarrow{P'P''} \perp L_2$. So, $\overleftrightarrow{PP'} = \overleftrightarrow{P'P''}$. **b.** As the distance between L_1 and L_2 decreases, the size of the translation decreases. When $L_1 = L_2$, every point is its own image. This may be thought of as a zero translation.

12. The successive reflections in parallel mirrors produce the same result as a translation, and the object appears to be in the direct line of vision.

13. A glide preserves orientation, and a glide reflection reverses orientation.

14. A translation

Section 1.9, Pages 39–41

1. Point symmetry: a, b, c, d, e, f; Line symmetry: a, b, c, f; Rotational symmetry: a, b, c, d, e, f

2. a. Line symmetry; $x = 0$ **b.** Line symmetry; $x = 0$ **c.** Line symmetry; $y = 0$ **d.** Line symmetry; $x = 0$ **e.** Line symmetry; $x = 0$ **f.** Line symmetry; $y = 0$ **g.** Point symmetry; $(0, 0)$ **h.** Point symmetry; $(0, 2)$ **i.** Point symmetry; $(0, 0)$

3. Point symmetry: H, I, N, O. S, X, Z; Line symmetry: A, B, C, D, E, H, I, M. O, T, U, V, W, X

4. a. 12 **b.** 6 **c.** 10 **d.** 5 **e.** 2

5. a. The word "bilateral" means two sides, and an object with bilateral symmetry has two congruent sides. **b.** All have bilateral symmetry. **c.** A jelly fish has point symmetry and rotational symmetry. An octopus, if the eyes and mouth are disregarded, has point symmetry and rotational symmetry.

6. A parallelogram has symmetries H_0 and I, where O is the center of the parallelogram. A rhombus has symmetries R_1, R_2, H_0, and I, where L_1 and L_2 are the lines containing its diagonals, and O is its center. A square has symmetries R_1, R_2, R_3, R_4, H_0, O_{90}, O_{270}, and I, where L_1 and L_2 are the lines containing its diagonals, L_3 and L_4 are the perpendicular bisectors of its sides, and O is its center.

7. If each figure were circumscribed by a square, the

symmetries of each figure would be precisely those eight symmetries of a square mentioned in Exercise 6.

8. a. $\overline{DE}$; a half-turn is an isometry. **b.** $\angle DEB$; an isometry preserves angle measure, and $\angle ABE$ and $\angle DEB$ are alternate interior angles.

Section 1.10, Page 44.
1. Only if the point and its image are distinct.
2. Yes
4. One; two

Section 2.1, Pages 51–54.
1. G
3. FM: maternal grandfather; FF: paternal grandfather; MF: paternal grandmother; MM: maternal grandmother
4. $R_y:(x, y) \rightarrow (-x, y)$ and $R_x:(-x, y) \rightarrow (-x, -y)$. $R_x:(x, y) \rightarrow (x, -y)$ and $R_y:(x, -y) \rightarrow (-x, -y)$. $R_x R_y = R_y R_x = H_o$.
5. a. No **b.** Yes **c.** No
8. FG$(A) = (-3, 1)$, FG$(B) = (3, 13)$, and FG$(C) = (7, 5)$. GF$(A) = (-6, 2)$, GF$(B) = (0, 14)$, and GF$(C) = (4, 6)$.
9. All are the same as the identity.
10. a. Yes **b.** No **c.** Yes **d.** No **e.** Yes **f.** No **g.** Yes
11. a. No **b.** No **c.** Yes **d.** No **e.** No **f.** No **g.** No
12. a. $-5, \frac{1}{5}$ **b.** $7, -\frac{1}{7}$ **c.** $-1, 1$ **d.** 0, zero has no multiplicative inverse. **e.** $-\frac{2}{5}, \frac{5}{2}$ **f.** $\frac{1}{2}, -2$
13. a. C_{-135} or C_{225} **b.** H_A **c.** $R_2 R_1$ **d.** I **e.** R_L **f.** GF
14. a. $(H_A H_B)(H_B H_A) = H_A(H_B H_B)H_A = H_A I H_A = H_A H_A = I$ **b.** The inverse of H_A is H_A and the inverse of H_B is H_B. So, the inverse of $H_A H_B$ is $H_B H_A$, by Theorem 14.
15. a. Both have the image X. **b.** Every image point has infinitely many preimages.
16. a. Reflection; inverse is F **b.** Half-turn; inverse is F **c.** Rotation; inverse is $(x, y) \rightarrow (y, -x)$ **d.** Translation; inverse is $(x, y) \rightarrow (x, y - 3)$ **e.** Translation; inverse is $(x, y) \rightarrow (x - 10, y)$ **f.** Translation; inverse is $(x, y) \rightarrow (x - 10, y - 3)$
17. Definition of composite; Substitution; Theorem 13; A reflection is its own inverse; Definition of identity
18. $R_3 R_2 R_1 = R_3(R_2 R_1) = R_3(R_3 R_4) = (R_3 R_3)R_4 = I R_4 = R_4$
19. *3.* $R_6 R_5$ is a glide by Theorem 8. L_7 is parallel to the direction of the glide, so by the definition of glide reflection, $R_7(R_6 R_5)$ is a glide reflection. *4.* Definition of composite; Substitution; Theorem 13; By Theorem 18, $R_3 R_4 = H_B$ and $R_7 R_6 = H_B$, and substitution; Theorem 13
20. a. $(2, -9)$ **b.** $(-1, -11)$ **c.** $(-6, -10)$ **d.** $(x, y - 10)$
21. a. $(2, 11)$ **b.** $(-1, 9)$ **c.** $(-6, 10)$ **d.** $(x, y + 10)$

Section 2.2, Pages 58–61.
1. c
2. a. $\overrightarrow{DC}$ **b.** $\overrightarrow{AD}$ **c.** $\overrightarrow{BO}$ **d.** $\overrightarrow{CO}$ **e.** $\overrightarrow{AC}$ **f.** $\overrightarrow{AC}$

3. a. $\overrightarrow{AD}, \overrightarrow{FE}$ **b.** $\overrightarrow{DC}$ **c.** $\overrightarrow{CE}$ **d.** $\overrightarrow{AE}$ **e.** $\overrightarrow{FC}$ **f.** $\overrightarrow{AE}$
5. $D(7, 3)$; $E(10, 7)$; $F(1, -5)$
6. a. False **b.** True **c.** True
7. a. $\overrightarrow{AC}$ **b.** $\overrightarrow{EC}$ **c.** $\overrightarrow{EC}$ **d.** $\overrightarrow{CD}$ **e.** $\overrightarrow{CD}$
8. a. $\overrightarrow{AC}$ **b.** $\overrightarrow{CB}$ **c.** $\overrightarrow{BC}$
9. a. $\overrightarrow{DC}, \overrightarrow{EF}, \overrightarrow{HG}$ **b.** $\overrightarrow{HE}, \overrightarrow{DA}, \overrightarrow{CB}$ **c.** $\overrightarrow{AH}$ **d.** $\overrightarrow{EG}, \overrightarrow{AC}$ **e.** $\overrightarrow{EC}$ **f.** $\overrightarrow{AG}$
10. a. $\overrightarrow{FD}$ **b.** $\overrightarrow{AG}$ **c.** $\overrightarrow{EC}$ **d.** $\vec{0}$ **e.** Identity translation or zero translation
11. $100\sqrt{17} \approx 412$ miles per hour; approximately east-southeast
12. a. $AC = \sqrt{26}$ and $BD = \sqrt{26}$, so $AC = BD$. Also, slope of $\overleftrightarrow{AC} = \frac{1}{5}$ and slope of $\overleftrightarrow{BD} = \frac{1}{5}$, therefore $\overline{AC} \| \overline{BD}$. Thus, $ABCD$ is a parallelogram. **b.** $(a + c, b + d)$
13. a. $(12, 8)$ **b.** $(18, 12)$ **c.** $(-6, -4)$ **d.** $(-12, -8)$
14. Approximately 51.38 miles; approximately south-southeast
17. a. $\vec{x} + \vec{y}$ **b.** $\vec{z} - \vec{x}$ **c.** $\vec{y} + \vec{z}$ **d.** $\vec{y} + \vec{z}$ **e.** $-\vec{x} - \vec{y} - \vec{z}$
18. b. $Q(3, 2)$; $R(3, -1)$; $S(-2, 1)$
19. Approximately 261 miles per hour; approximately north-northwest
20. 5 miles per hour; south-southwest
21. By the definition of translation, $ABQP$ and $BCRQ$ are parallelograms. Therefore, $AP = BQ$, $\overline{AP} \| \overline{BQ}$, $BQ = CR$, and $\overline{BQ} \| \overline{CR}$. Both equality and parallelism are transitive, so $AP = CR$ and $\overline{AP} \| \overline{CR}$. Therefore, $ACRP$ is a parallelogram and $\overrightarrow{AC} = \overrightarrow{PR}$.
22. Let D be the point such that $ABCD$ is a parallelogram. Then, $\overrightarrow{AB} = \overrightarrow{DC}$, and $\overrightarrow{AD} = \overrightarrow{BC}$. Translation S maps A to D and translation T maps D to C. Therefore, TS maps A to C. Since A was an arbitrary point and $ST(A) = TS(A)$, then $ST = TS$.
23. Let $H_A(P) = P^*$. Consider the triangle PP^*P'. By the definition of half-turn, A is the midpoint of $\overline{PP^*}$ and B is the midpoint of $\overline{P^*P'}$. Therefore, $AB = \frac{1}{2}PP'$ and $\overline{AB} \| \overline{PP'}$. Since A is in $\overline{PP^*}$ and B is in $\overline{P'P^*}$, the direction from A to B is the direction from P to P'. Thus, $\overrightarrow{PP'} = 2\overrightarrow{AB}$.
24. If $\vec{a}$ and $\vec{b}$ are perpendicular and both are one unit long, then the figure becomes a rectangular coordinate system.

Section 2.3, Pages 66–67.
3. b. Infinitely many
4. a. If $\overrightarrow{AB} = \overrightarrow{CD}$. **b.** $(H_D H_C)(H_C H_D) = H_D(H_C H_C)H_D = H_D I H_D = H_D H_D = I$. Therefore, $I = H_D H_C H_B H_A$.

5. The composite of five half-turns is a half-turn, by Theorem 21. The center of this half-turn would have to be the midpoint of $\overline{PP'}$. By the definition of a half-turn, this point would also be the midpoint of $\overline{QQ'}$.

6. **a.** The composite of five half-turns is a half-turn by Theorem 21. Let $H_E H_D H_C H_B H_A = H_X$. Then, $(H_E H_D H_C H_B H_A)^2 = (H_X)^2 = I$, since a half-turn is its own inverse.

7. $H_E H_D H_C H_B H_A$ will be equivalent to some half-turn H_Y. We want $H_Y(P) = P$, therefore, $Y = P$. That is, P must be the center of the half-turn. For a point X with image X', P will be the midpoint of $\overline{XX'}$. To complete the pentagon, find $H_A(P) = Q$, $H_B H_A(P) = R$, $H_C H_B H_A(P) = S$, and $H_D H_C H_B H_A(P) = T$.

8. $TS = (H_Q H_B)(H_B H_P) =$
$H_Q(H_B H_B)H_P = H_Q I H_P = H_Q H_P$.

9. $H_A = R_2 R_1$ and $H_B = R_1 R_3$ by Theorem 18. Therefore, $H_A R_3 = R_2 R_1 R_3$ and $R_2 H_B = R_2 R_1 R_3$. Therefore, $H_A R_3 = R_2 H_B$.

10. Consider the composite of n half-turns, when n is an even number. We may group these half-turns in pairs. The composite of the first and the second is equivalent to a translation, the composite of the third and fourth is equivalent to a translation, and so on. This product of n half-turns will be equivalent to the product of $\frac{n}{2}$ translations. By Theorem 15, the composite of translations is a translation. Thus, when n is even, the product of n half-turns is a translation.

When n is an odd number, grouping in pairs will result in the product of $\frac{n-1}{2}$ translations and a half-turn. By Theorem 15 and the Corollary to Theorem 20, this is equivalent to a half-turn.

11. By Theorem 19, $H_B H_A = 2\overrightarrow{AB}$. Therefore, if M is the midpoint of $\overline{AB}$, $H_M H_A = 2\overrightarrow{AM}$ and $H_B H_M = 2\overrightarrow{MB}$. Since $\overrightarrow{AM} = \frac{1}{2}\overrightarrow{AB}$ and $\overrightarrow{MB} = \frac{1}{2}\overrightarrow{AB}$, $H_M H_A = \overrightarrow{AB}$, and $H_B H_M = \overrightarrow{AB}$.

12. If $H_B H_A = H_C H_D$, then $\overrightarrow{AB} = \overrightarrow{DC}$. That is, $AB = DC$ and $\overline{AB} \| \overline{DC}$. Therefore, $ABCD$ is a parallelogram.

If $ABCD$ is a parallelogram, $AB = DC$ and $\overline{AB} \| \overline{DC}$. Since the direction from A to B is the same as the direction from D to C, $\overrightarrow{AB} = \overrightarrow{DC}$. Therefore, $2\overrightarrow{AB} = 2\overrightarrow{DC}$, and by Theorem 19, $H_B H_A = H_C H_D$.

13. **b.** $H_A R_1$ is equivalent to a reflection in L_2, the line perpendicular to L_1 at A. If $H_A = R_2 R_1$, then $H_A R_1 = R_2 R_1 R_1 = R_2 I = R_2$.

14. By Theorem 18, if $L_2 \perp L_3$ at A, then $H_A = R_3 R_2$. Therefore, $H_A R_1 = R_3 R_2 R_1$. If L_2 and L_3 have been situated so that $L_1 \| L_2$, then $R_2 R_1$ is a translation by

Theorem 8. The direction of the translation is perpendicular to L_2 and L_1, and therefore, is parallel to L_3. By definition, $R_3 R_2 R_1$ is a glide reflection.

15. $H_C:(x, y) \to (2c - x, -y)$, $H_B:(2c - x, -y) \to (x - 2c, y)$, and $H_A:(x - 2c, y) \to (2c - x, 2a - y)$. This product has the same effect as a half-turn about the point (c, a) which completes the parallelogram.

16. Let A and B be the two fixed points. Consider point X, an arbitrary point of the plane which is not contained in $\overleftrightarrow{AB}$. Let X' be the image of X by the given isometry. If $X \neq X'$, then $AX = AX'$ and $BX = BX'$, so $\overleftrightarrow{AB}$ is the perpendicular bisector of $\overline{XX'}$. Therefore, the isometry is a reflection in $\overleftrightarrow{AB}$.

If $X = X'$, let us consider what the isometry does to a fourth point, Y. Assume the image of Y is Y'. Then, A, B, and X all lie in the perpendicular bisector of $\overline{YY'}$. This cannot be true since A, B, and X are noncollinear. Therefore, $Y = Y'$ so the isometry must be the identity.

Section 2.4, Page 70.
1. **a.** A_{150} **b.** A_{-8} **c.** C_{270} **d.** $C_{360} = I$
2. b, c, d, and e are all translations.
3. $T = R_2 R_1$ by Theorem 8, and $A_x = R_1 R_3$ by Theorem 6.
$TA_x = (R_2 R_1)(R_1 R_3) = R_2(R_1 R_1)R_3 = R_2 I R_3 = R_2 R_3$.
By Theorem 6, $R_2 R_3$ is a rotation about point C through twice the angle of rotation from L_3 to L_2.
Since $L_1 \| L_2$, this angle is $\frac{x°}{2}$, and $R_2 R_3 = C_x$.

Section 2.5, Pages 74–75.
1. **a.** Yes **b.** No **c.** Yes **d.** No **e.** No **f.** No
   g. Yes **h.** No
2. **a.** Yes **b.** Yes **c.** No **d.** Yes **e.** No **f.** Yes
3. $b \circ d = e$ and $d \circ b = f$
4. **a.** Closed under addition **b.** Closed under multiplication **c.** Closed under addition, subtraction, and multiplication **d.** Closed under multiplication and division **e.** Closed under addition, subtraction, and multiplication
5. Each element is the same as the element in its column heading. Special columns are headed by 0 in table a and by c in table b.
6. **a.** d **b.** a and c are inverses; b and d are their own inverses. **c.** a **d.** Each element is its own inverse.
7. The set of integers is closed under addition. Addition is an associative operation. The additive identity of the set of integers is 0, and the additive inverse of an integer a is $-a$, which is also an integer. Thus, the set of integers is a group with addition.

To show that the set of integers is not a group with multiplication note that not every integer has a multiplicative inverse which is an integer.
8. Yes; No, because the odd integers are not closed under addition.

9. a. *1.* Any two elements of this set can be written in the form $3m$ and $3n$, where m and n are integers. Therefore, $3m + 3n = 3(m + n)$. Since $m + n$ is an integer, the set is closed under addition. *2.* Since addition of integers is associative, this set has the associative property of addition. *3.* Zero is a multiple of 3, so the set has an identity element. *4.* Any element of the set can be written in the form $3m$. The inverse of $3m$ is $-3m$ or $3(-m)$, which is a multiple of 3. **b.** Yes

10.

$*$	a	b	c
a	c	a	b
b	a	b	c
c	b	c	a

b is the identity element. a and c are inverses; b is its own inverse.

11. a. Since $a \neq 0$, a has an inverse, $\dfrac{1}{a}$. $ab = ac$;

$\dfrac{1}{a}(ab) = \dfrac{1}{a}(ac)$; $\left(\dfrac{1}{a} \cdot a\right)b = \left(\dfrac{1}{a} \cdot a\right)c$; $1 \cdot b = 1 \cdot c$;

$b = c$. **b.** $H_A H_B = H_A H_C$; $H_A(H_A H_B) = H_A(H_A H_C)$; $(H_A H_A)H_B = (H_A H_A)H_C$; $IH_B = IH_C$; $H_B = H_C$. **c.** The set has an identity element, i, and a has an inverse a^{-1}. $a * b = a * c$; $a^{-1} * (a * b) = a^{-1} * (a * c)$; $(a^{-1} * a) * b = (a^{-1} * a) * c$; $i * b = i * c$; $b = c$. **d.** Assume d is repeated in the a-row under the columns headed b and c. Then, $a * b = d$ and $a * c = d$. That is, $a * b = a * c$. This implies $b = c$, which is impossible.

Section 2.6, Page 79.
1. By Theorem 22, the set of rotations with the same center is closed under composition. By Theorem 13, composition of rotations with the same center is associative. The identity element of this set is the zero rotation, and if C_z is an element of this set, its inverse C_{-z} is also in the set.
3. a. $D_3 D_2 : (x, y) \rightarrow (6x, 6y)$; $D_2 D_3 : (x, y) \rightarrow (6x, 6y)$.
b. $D_2 D_{\frac{1}{2}} : (x, y) \rightarrow (x, y)$; D_1 **c.** Let S be the set of all dilations with center O. S is closed because if D_m and D_n are in S, then $D_m D_n = D_{mn}$, which is also in S. Composition is always associative, and in this case, it is always commutative, since $D_m D_n = D_{mn} = D_{nm} = D_n D_m$. The identity transformation is D_1, which is in S. Finally, the inverse of D_m is $D_{1/m}$, which is also in S.
4. Let F and G be any two isometries. If $F(A) = A'$ and $F(B) = B'$, then $AB = A'B'$. Also, if $G(A') = A''$ and $G(B') = B''$, then $A'B' = A''B''$. Since equality is transitive, $AB = A''B''$ and GF preserves distance.
5. a. Since $\overrightarrow{OP'} = 2\overrightarrow{OP}$, P is the midpoint of $\overline{OP'}$. Also, $\overrightarrow{OQ'} = 2\overrightarrow{OQ}$ so Q is the midpoint of $\overline{OQ'}$. Therefore, $\overline{PQ} \| \overline{P'Q'}$. **b.** P, Q, and R are the midpoints of $\overline{OP'}$, $\overline{OQ'}$, and $\overline{OR'}$, respectively. There-

fore, $PQ = \frac{1}{2}P'Q'$; $QR = \frac{1}{2}Q'R'$, and $RP = \frac{1}{2}R'P'$. $\triangle PQR \sim \triangle P'Q'R'$, thus, $m\angle PQR = m\angle P'Q'R'$.
c. $PQ = \frac{1}{2}P'Q'$ and $QR = \frac{1}{2}Q'R'$, so

$$\frac{PQ}{QR} = \frac{\frac{1}{2}P'Q'}{\frac{1}{2}Q'R'} = \frac{P'Q'}{Q'R'}.$$

6. a. The midpoint of $\overline{AB}$ is $(1, \frac{9}{2})$. Its image by M is $(2, \frac{3}{2})$. But, the midpoint of $\overline{A'B'}$ is $(2, \frac{3}{2})$. Use the same procedure for $\overline{BC}$ and $\overline{AC}$. **b.** Because it maps A, B, and C to A', B', and C' and it maps the midpoints of $\overline{AB}$, $\overline{BC}$, and $\overline{AC}$ to the midpoints of $\overline{A'B'}$, $\overline{B'C'}$, and $\overline{A'C'}$. **c.** The centroid of $\triangle ABC$ belongs to all three medians. By part b, the image of the centroid must belong to all three medians of $\triangle A'B'C'$ and so must be the centroid of $\triangle A'B'C'$.
7. a, c, i, and j are equivalent (no holes). b and e are equivalent (one hole). f and g are equivalent (two holes). d and h are equivalent (three holes).

Section 2.7, Pages 83–84.
1. These symmetries form a commutative group whose table is identical to the one shown on page 80.

3.

	I	O_{120}	O_{240}	R_1	R_2	R_3
I	I	O_{120}	O_{240}	R_1	R_2	R_3
O_{120}	O_{120}	O_{240}	I	R_3	R_1	R_2
O_{240}	O_{240}	I	O_{120}	R_2	R_3	R_1
R_1	R_1	R_2	R_3	I	O_{120}	O_{240}
R_2	R_2	R_3	R_1	O_{240}	I	O_{120}
R_3	R_3	R_1	R_2	O_{120}	O_{240}	I

4. In square $ABCD$, let L_1 be diagonal $\overline{AC}$, let L_2 be diagonal $\overline{BD}$, let L_3 be the perpendicular bisector of $\overline{AD}$ and $\overline{BC}$, let L_4 be the perpendicular bisector of $\overline{AB}$ and $\overline{DC}$, and let O be the center of the square.

	I	O_{90}	O_{180}	O_{270}	R_1	R_2	R_3	R_4
I	I	O_{90}	O_{180}	O_{270}	R_1	R_2	R_3	R_4
O_{90}	O_{90}	O_{180}	O_{270}	I	R_3	R_4	R_2	R_1
O_{180}	O_{180}	O_{270}	I	O_{90}	R_2	R_1	R_4	R_3
O_{270}	O_{270}	I	O_{90}	O_{180}	R_4	R_3	R_1	R_2
R_1	R_1	R_4	R_2	R_3	I	O_{180}	O_{270}	O_{90}
R_2	R_2	R_3	R_1	R_4	O_{180}	I	O_{90}	O_{270}
R_3	R_3	R_1	R_4	R_2	O_{90}	O_{270}	I	O_{180}
R_4	R_4	R_2	R_3	R_1	O_{270}	O_{90}	O_{180}	I

Index

Affine transformation 78
Angle of incidence 17
Angle of reflection 17
Angle of rotation 26
Associativity 48, 49, 71

Closure 70–71
Commutative group 72
Commutativity 47, 72
Composite 46
Composition 46, 76

Dilation 77
Domain 5

Escher, M. C. 36, 45, 83

Fixed point 4, 49
Functions 9, 10

Glide (*see* translation)
Glide reflection 33–34
Group 70–74, 76–78, 80–83
    commutative 72
    definition of a 71
    symmetry 80–83
    transformation 76–78

Half-turn(s) 13, 62–66
    as a composite of reflections 62
    composite of 63, 64, 65
    inverse of a 49

Identity element 49, 71
Identity mapping 27, 35, 38
Image 2, 5
Inverses 10, 50, 51, 71
Isometry(ies)
    as a group 76, 78
    definition of 12
    Fundamental Theorems of 41–44
    properties of 22–24

Latitude, circles of 3
Line reflection(s) 12–13
    composite of 28, 33, 43–44, 62
    definition of 12
    inverse of a 49
Line symmetry 37
Longitude, circles of 3

Mappings 3, 5–7, 9, 10
    one-to-one 6, 9
Maps 2, 3, 4, 5
Meridian 3

One-to-one mapping 6, 9
Orientation 24, 28, 35

Parallels 3
Point reflection (*also see* half-turn)
    definition 13
Point symmetry 37
Preimage 5
Preservation of
    angle measure 23
    area 4
    distance 3, 4
    orientation 24, 28, 35
    parallelism 23
Product of transformations 46
    inverse of a 51
Projection of the plane onto a line 6
Pythagorean Theorem 30

Range 5
Reflection in a line (*see* line reflection)
Reflection in a point (*see* point reflection and
    half-turn)
Rotation(s) 26–29, 68–70
    angle of 26
    as a composite of line reflections 28
    composite of 68–69
    definition of 27
    inverse of 50
    zero rotation 27
Rotational symmetry 38

Scale factor 77
Similarity transformation 77, 78
Symmetry 37–38
   definition of 38
Symmetry group 80–83

Topological transformation 77, 78
Transformation(s) 7
   affine 78
   composite of 46, 62
   group 76–78
   similarity 77, 78
   topological 77, 78

Translation(s) 31–33, 55–58
   as a composite of reflections 33
   composite of 58
   definition of 32
   inverse of 50
   represented by an arrow (vector) 55
   represented by an ordered pair 60
   zero translation 35

Vector(s)
   parallelogram method of addition 56
   representing a translation 55
   triangle method of addition 57